IFCoLog Journal of Logics and their Applications

Volume 3, Number 3

September 2016

Disclaimer

Statements of fact and opinion in the articles in IfCoLog Journal of Logics and their Applications are those of the respective authors and contributors and not of the IfCoLog Journal of Logics and their Applications or of College Publications. Neither College Publications nor the IfCoLog Journal of Logics and their Applications make any representation, express or implied, in respect of the accuracy of the material in this journal and cannot accept any legal responsibility or liability for any errors or omissions that may be made. The reader should make his/her own evaluation as to the appropriateness or otherwise of any experimental technique described.

ISBN 978-1-84890-222-0
ISSN (E) 2055-3714
ISSN (P) 2055-3706

College Publications
Scientific Director: Dov Gabbay
Managing Director: Jane Spurr

http://www.collegepublications.co.uk

Printed by Lightning Source, Milton Keynes, UK

SCOPE AND SUBMISSIONS

This journal considers submission in all areas of pure and applied logic, including:

pure logical systems
proof theory
constructive logic
categorical logic
modal and temporal logic
model theory
recursion theory
type theory
nominal theory
nonclassical logics
nonmonotonic logic
numerical and uncertainty reasoning
logic and AI
foundations of logic programming
belief revision
systems of knowledge and belief
logics and semantics of programming
specification and verification
agent theory
databases

dynamic logic
quantum logic
algebraic logic
logic and cognition
probabilistic logic
logic and networks
neuro-logical systems
complexity
argumentation theory
logic and computation
logic and language
logic engineering
knowledge-based systems
automated reasoning
knowledge representation
logic in hardware and VLSI
natural language
concurrent computation
planning

This journal will also consider papers on the application of logic in other subject areas: philosophy, cognitive science, physics etc. provided they have some formal content.

Submissions should be sent to Jane Spurr (jane.spurr@kcl.ac.uk) as a pdf file, preferably compiled in LaTeX using the IFCoLog class file.

CONTENTS

Editorial Preface 279
    Heinrich Wansing, Hitoshi Omori, and Thomas Macaulay Ferguson

ARTICLES

Connexive Implication, Modal Logic and Subjunctive Conditionals 297
    Richard Bradshaw Angell

Models for Connexive Logics 309
    Richard Routley and Hugh Montgomery

On a New Three-Valued Paraconsistent Logic 317
    Grigory Olkhovikov

A Complete, Correct, and Independent Axiomatization of the First-Order
    Fragment of a Three-Valued Paraconsistent Logic 335
    Grigory Olkhovikov

A Comparison of Connexive Logics 341
    Luis Estrada-González and Elisángela Ramírez-Cámara

On Arithmetic Formulated Connexively 357
    Thomas Macaulay Ferguson

Beyond System P – Hilbert-Style Convergence Results for Conditional
Logics with a Connexive Twist 377
Matthias Unterhuber

Natural Deduction for Bi-Connexive Logic and a
Two-Sorted Typed λ-Calculus 413
Heinrich Wansing

Completeness of Connexive Heyting-Brouwer Logic 441
Norihiro Kamide and Heinrich Wansing

A Simple Connexive Extension of the Basic Relevant Logic BD 467
Hitoshi Omori

Natural Deduction for Two Connexive Logics 479
Nissim Francez

A Note on Francez' Half-Connexive Formula 505
Hitoshi Omori

The Tenacity of Connexive Logic: Preface to the Special Issue

Heinrich Wansing
Department of Philosophy II
Ruhr-University Bochum, Germany
Heinrich.Wansing@rub.de

Hitoshi Omori
Department of Philosophy
Kyoto University, Japan
hitoshiomori@gmail.com

Thomas Macaulay Ferguson
Department of Philosophy
City University of New York Graduate Center, United States
tferguson@gradcenter.cuny.edu

1   Introduction

If one's introduction to philosophical logic has been carried out in the shadow of the canonical history of Western logic, one may be forgiven for believing that connexive theses concerning reasoning—*e.g.*, that no proposition entails or is entailed by its own negation—have been soundly defeated in favor of their classically-accepted cousins. Even among work devoted to examining non-classical themes, the discussion of connexive principles frequently either warrants only a few negative comments (*e.g*, the Kneales' [15]) or is omitted entirely (*e.g.*, Haack's [12], Bell, de Vidi, and Solomon's [6], and Priest's [36]).[1] Among logicians and philosophers who are generally sympathetic to—or tolerant of—non-classical themes, connexive logic has been given remarkably short shrift.

---

[1]In fairness, Priest's second edition [37] includes a brief discussion of the connexive logic introduced in [50]. Another exception is Humberstone's [13].

Hence, the warrant for the present collection—its legitimacy—is intimately tied to the question: What explains the status of connexive logic in the current landscape?

The historical appearances of such theses are often treated as tokens of a common mistake made by otherwise brilliant thinkers. But when one surveys the myriad occasions in which such "mistakes" were committed—and the brilliance of their respective transgressors—it becomes increasingly difficult to dismiss these instances as errors. Aristotle appeals to connexive principles in the *Organon* both explicitly and implicitly. It is clear from the studies of syllogistic and the square of opposition in Hugh MacColl's [21], Storrs McCall's [26], and Luis Estrada-González' [9] that there exists a deep connection between Aristotelian syllogistic and connexive implication (a fact that features heavily in Estrada-González and Elisángela Ramírez-Cámara's contribution to this collection). In medieval and early modern studies of inference, such theses were also common; Priest's [35] and Richard Sylvan's [46] deftly document countless other appearances of these principles in the texts of philosophers ranging from Abelard to Berkeley.

But even in the era following Frege, the eradication of connexive theses has proven incredibly stubborn. Like a logical Zelig, when one traces many trends in logic to their sources, one finds connexive theses waiting. With respect to the modern treatment of modal logic, C.I. Lewis is frequently cited as its progenitor. Despite this, Lewis' formalizations of modal systems owe a great deal to the *Calculus of Equivalent Statements* developed by Hugh MacColl in [17, 18, 19, 20]. (Stephen Read suggests that MacColl's role was obscured by the fact that Lewis was "not fully candid... in acknowledging his debt to MacColl" [42, p. 59].) Insofar as MacColl's calculus admits connexive theorems in special cases—a fact discussed in detail in [22]—it is fair to say that connexive intuitions played a strong role in the incubation of modern modal logic. A similar connexive provenance holds of the field of conditional logic, whose themes are seeded in the *Ramsey test*—the method of evaluating conditionals described in Frank Ramsey's famous footnote in [40]. As observed in [10] and [27], however, Ramsey's account of conditionals assumes the validity of connexive principles, namely, Boethius' Thesis, a feature that rests at the heart of Matthias Unterhuber's contribution to this special issue. In these cases, we find a connexive specter haunting entire fields of non-classical logic.

Many of the obstacles to connexive logic's securing a foothold in the canon have been purely contingent. Some of these are largely political in nature. For example, while intuitionistic and relevant logics reaped the benefits of careful curation by their respective champions (*e.g.*, Anderson and Belnap's [1]), connexive logic has not enjoyed a similar benefactor. The lack of a unified *vade mecum* has led to problems that would cripple the dissemination of any formal system. In many ways, this has had a devastating effect on the very coherence of the term "connexive logic"; while careful curation ensured that important non-classical notions like constructivity and relevance received relatively distinct and salient definitions, the very concept of a "connexive logic" remains quite fluid. The appearance of sources such as [52] and [27] has undoubtedly served to ameliorate this deficit, but its effects still linger.

Some impediments, however, are more entrenched. Frequently, those best positioned to take up the mantle of champion of connexive logic—like Everett Nelson or Storrs McCall—quickly disavowed the particular implementations of connexive principles they had introduced. We might consider the collection of work on connexive logic scattered between the early 1960s and the late 1980s to constitute the "first wave" of the modern era of connexive logic.

Judging by the fruits of these earlier papers, it is not entirely unreasonable to greet the notion of connexive logic with some suspicion. The semantics offered for connexive systems during this period did not arise from any salient portraits of familiar semantic notions, but were offered strictly as tools in the service of *proof-theoretic* properties about the favored axiom systems. As a result, authors during this period explicitly postponed any discussion of salient or useful interpretations for their semantics (*e.g.*, [2] and [24]). From the axiomatic side, too, the costs of adopting Aristotle and Boethius' Theses appeared to be astonishingly severe, frequently requiring the rejection of one or more *prima facie* valid inferences. Among the early systems of connexive logic, one of the most suspicious sacrifices was the rejection of conjunctive simplification—the inference from a conjunction to each of its conjuncts—as observed, *e.g.*, by John Woods in [55].

Problematically, on the few occasions in which logicians *did* defend these sacrifices on philosophical grounds, the defenses tended towards opaqueness. In the particular case of the invalidity of conjunctive simplification, the arguments Everett Nelson outlines against the inference in, *e.g.*, [29] and [30] are extraordinarily esoteric. Furthermore, Bruce Thompson's arguments in [47] center around a notion of "reversal negation" that is primarily motivated by a vague and imprecise analogy of laying down, removing, rotating, and tearing apart playing cards on a table.

But in hindsight, some of these discussions become more sympathetic. While McCall and Angell's many-valued semantics were often decried as counterintuitive, it has become far more common to divorce the utility of a truth-functional semantics from an intuitive interpretation (primarily due to the success of matrix semantics in the study of modal and relevant logics). Furthermore, some of the discussion might be viewed more favorably in the light of contemporary developments in logic. For example, Everett Nelson in [29] rejects the inference from $\varphi \wedge \psi$ to φ because ψ need not have been "used" to derive the conclusion. Insofar as simplification fails for multiplicative conjunction in *linear logic* for precisely the reasons Nelson gives in [29]—that is, because not all resources identified in the premises are being *used* in the derivation—one might argue that Nelson is an unsung herald of linear logic. When one removes "reversal negation" from the picture in Thompson's [47], too, some of its arguments become more convincing. Thompson, for example, states that we must "guard against" formulae such as:

- $(p \land \neg(p \to p)) \to p$

This formula is read by Thompson as the assertion that "'p' follows from 'p', even under the condition that it does not."[47, p. 253] The orthodoxy explains that p is a consequence of one of the conjuncts—in this case, the first conjunct—because p is a consequence of itself. The validity of conjunctive simplification, then, presupposes the validity of self-implication, an inference that itself has been challenged in other contexts, such as Errol Martin and Robert Meyer's S in [23].

For many reasons, we suggest that connexive logic has been too casually rejected.

2   The Persistence of Connexive Principles

To address the matter of the merits of a study of connexive logic, it helps to consider the matter of the sheer persistence and tenacity of its primary theses. Much of their persistence in philosophical logic can be considered a consequence of a simple feature that holds in classical propositional logic—and hence, in each of its fragments and subsystems (*e.g.*, intuitionistic logic): The only counterexamples to connexive theses such Aristotle and Boethius' Theses are those in which the antecedent of the conditional is *logically impossible*. In (for example) [11], we find the following observation:

Observation. *For a formula φ, if φ is satisfiable in classical logic, then $\neg\varphi$ is not a classical consequence of φ.*

In other words, it is *only* contradictions that entail their own negations. Consequently, because nearly all common non-classical propositional logics remain *fragments* of classical logic, this fact holds of constructive logics, logics of strict implication, relevant logics, fuzzy logics, *etc.* as well.

This fact has a number of consequences for the tenacity of connexive principles: Firstly, while there are infinitely many *formulae* that are classically inconsistent, the most entrenched philosophical accounts of *meaning* suggest that all of these represent a *single* proposition, entailing a scarcity of propositions with respect to which Aristotle's Thesis (AT) fails. Secondly, the most common accounts of semantic validity involve the preservation of some property—truth, non-falsity, meaningfulness—across all models. In the case of classical logic and its non-paraconsistent subsystems, contradictions have no models, whence the validity of all counterexamples to Aristotle's Thesis is *vacuously* satisfied.

2.1 The Paucity of Counterexamples

In his contribution to this special issue, Matthias Unterhuber describes Aristotle's Thesis by means of an incredibly apt and perceptive phrase: Aristotle's Thesis, says Unterhuber, is "almost true." Appropriately understood, this description is apt as the property of being "almost true" indeed holds of Aristotle's Thesis in, *e.g.*, classical logic in a very robust and precise sense. Moreover, the appropriateness of this description of AT parallels one of the reasons for the tenacity of connexive logic, namely, the sheer paucity of counterexamples to the archetypal connexive theses.

In a formal setting, the term "almost all" has a very precise meaning. If one fixes two infinite sets S and T such that $S \subseteq T$, to say that "almost all Ts are Ss" (or, "almost all members of T are members of S") is to say that S is *cofinite* in T, that is, that the infinite set S includes all but a finite number of elements of T. Such situations become more interesting when probability is taken into account. That *almost all* elements of T are elements of S suggests that an arbitrary element $a \in T$ will *almost always* be a member of S, that is, that for an arbitrary $a \in T$, one may infer that a is a member of S as well *with probability 1*. That T is infinite while the relative complement $T \setminus S$ is finite means the probability of selecting an element *not in S* is smaller than any positive, real-valued probability.

To bring this to bear on Aristotle's Thesis, recall that the counterexamples to AT in classical logic are precisely those formulae $\varphi \to \neg\varphi$ where φ is a contradiction. That AT can be considered "almost true" is underscored by the fact that most theories of *propositions*—that is, the *meanings* expressed by sentences—support the notion that *almost all* propositions do not entail their own negations.

For example, in the frequently-encountered possible-worlds interpretation of propositions, the proposition expressed by a sentence φ is identified with the set of possible worlds at which φ is true. Now, because contradictions are true at no world, in this type of model, all contradictions express the same proposition, *i.e.*, the empty set.[2] On this interpretation of propositions, then, that there is a single inconsistent proposition entails that *there is a single proposition* that entails its own negation. From this perspective, we face a picture in which there exist infinitely many propositions that *do not* entail their own negations. Because there are only

---

[2]In some semantics, a single "absurd world," at which all sentences are true, is identified. This account agrees that all contradictions express the same proposition, differing only for exchanging the empty set for the singleton containing only the absurd world.

finitely many propositions that *do* entail their own negations, this can be strengthened to the claim that *almost all* propositions do not entail their own negations. Consequently, Aristotle's Thesis correctly describes the inferential status of *almost all* propositions. Aristotle's Thesis is quite literally *almost always* true.

This has a number of interesting consequences in the context of Niki Pfeifer's experimental study [34]. Pfeifer provided experimental evidence that under very reasonable interpretations of the conditional, respondents endorsed Aristotle's Thesis far more frequently than they rejected it as false. The set-up of Pfeifer's study presented participants with the sentence:

It is not the case, that: If not-A, then A.

and asked whether the sentence was "guaranteed to be false," "guaranteed to be true," or if one could not infer its truth or falsity on its face.

Now, participants responded that Aristotle's Thesis was "guaranteed to be true" roughly three times as frequently as it was rejected, a result which—to classical eyes—might be thought of as providing evidence of the irrationality of most students. Viewed through this lens, such an observation might be understood as similar to experimental results in which players do not act rationally from the perspective of game theory, that is, cases in which respondents inconsistently assign utility values to goods, or in which agents have inconsistent sets of beliefs. But the above considerations suggest that there is something markedly different about these types of case.

Can the disagreement with classical principles demonstrated by the subjects of Pfeifer's study be in any way thought *irrational*? If one should elect to follow Aristotle and refuse to employ any inference from a statement to its negation in one's reasoning, would one's welfare be crippled by adhering to this strategy?

The respondents were asked whether instances of Aristotle's Thesis were "*guaranteed* to be true," and it seems that interpreting the claim that "ϕ is guaranteed to be true" as "ϕ is true with probability 1" is an entirely natural one. Rather than being irrational, then, this reading *vindicates* the respondents who believed that Aristotle's Thesis was guaranteed. From a more pragmatic perspective, what would the consequences be for the layperson—that is, an individual who does not make use of formal proof in her day-to-day life—if she should accept Aristotle's Thesis as a correct principle, and employ it in her day-to-day reasoning? Quite possibly—quite likely, even—most individuals have never uttered a sentence for which Aristotle's Thesis will lead them astray. Following Aristotle, then, hardly seems irrational.

2.2 The Vacuity of Counterexamples

Beyond the poverty of the class of propositions for which Aristotle's Thesis fails, further consideration of the classical counterexamples to Aristotle's Thesis reveals an additional *vacuity* implicit in each of these cases. This vacuity emerges when one considers any counterexample to Aristotle—say, the validity of the inference from $p \wedge \neg p$ to $\neg(p \wedge \neg p)$—and asks *why* the inference is valid. From a classical perspective, the validity of the inference may be exhaustively explained by appeal to the vacuously true statement: Every model in which $p \wedge \neg p$ is true is one in which $\neg(p \wedge \neg p)$ is likewise true, something that holds precisely because there are *no* models satisfying $p \wedge \neg p$.

There seems to be an innate suspicion for such vacuous occasions of inference. The truth of the sentence "All unicorns have a horn" can be explained not on the basis of any property exhibited by unicorns, but solely on the basis of the non-existence of unicorns. This particular type of explanation plays a role in the distinction between the traditional and modern squares of opposition—and thus, modern syllogistic—and we have noted that [21], [26], and [9] tie these notions to connexive implication.

Intriguingly, the vacuity inherent in classical counterexamples to Aristotle shares important traits with Andreas Kapsner's critique of "empty promise conversions" in constructive semantics. In [14], Kapsner appeals to such vacuity when forming his arguments against the intuitionistic account of falsification. The Brouwer-Heyting-Kolmogorov (BHK) interpretation of the conditional suggests that a proof of a conditional $\varphi \to \psi$ is a computable function that converts any proof of φ into a proof of ψ. Consequently, a proof of an intuitionistically negated formula $\neg\varphi$—interpreted by Kapsner as an intuitionistic falsification of φ—converts any proof of φ into a proof of the absurdity $\bot$. Kapsner's critique of "empty promise conversions" may be seen as taking up earlier criticism of the BHK-clause for intuitionistic negation. In [49] it is observed that according to the BHK semantics, an intuitionistically negated formula $\neg A$ is valid if and only if there exists a construction that outputs a non existent object, namely a proof of $\bot$, when applied to a proof so A. This condition can be satisfied only vacuously for unprovable formulas A.

Ideally, the BHK-type interpretations of intuitionistic connectives are robust and non-vacuous. A proof of, say, $\varphi \to (\varphi \vee \psi)$ can be easily conceived as the process of taking a proof of φ and applying a single instance of disjunction introduction to its terminal line. In the cases in which we are interested, however—cases in which the antecedent of a conditional is logically impossible—the BHK clause can be satisfied in a much less satisfying way. As Kapsner observes, insofar as one will *never* be faced with an intuitionistic proof of a contradiction, in this case one may fulfill the letter of the BHK requirement by an "empty promise." One does not *need* to design or construct an appropriate algorithm. The logical impossibility of a formula φ precludes the existence of a proof of φ. Hence, the collection of all proofs of the formula φ is *empty, vacuously* entailing that one has a recursive method to convert all members of this collection into proofs of ψ.[3]

In instances of the vacuous satisfaction of the BHK clause by an "empty promise" —and, consequently, in those cases with respect to which Aristotle's Thesis fails— there is a fault in the *explanatory power* of the proof. On its face, the BHK interpretation promises something tangible and satisfying—in the case of a conditional $\varphi \to \psi$, one expects the description of a computation that *explains* how proofs of φ lead to proofs of ψ. To Kapsner, empty promises pose a problem for the intuitionistic account of falsity because there is no *requirement* that any work be done in demonstrating how $p \wedge \neg p$ comes to imply $\bot$. In the case of counterexamples to Aristotle, these vacuous conversions undermine the promise of BHK semantics because they do not require that a proof of the antecedent *causes, constructs,* or otherwise *brings about* a proof of the consequent, leading to a sort of *inferential inertness*.[4]

Applying the themes of Kapsner's [14] to the particular class of counterexamples to Aristotle's Thesis permits us to resituate the classical problems at the heart of connexive themes. For example, Kapsner's analysis can be brought to bear on the notion of *self-refutation* by appealing to the corollary that the BHK semantics permit all self-refutations to be conducted on the basis of empty promises. The theme of self-refutation lies at the heart of many the passages in which Aristotle invokes connexive themes. In Luca Castagnoli's work on the subject of self-refutation [8] we find several instances in which Aristotle's connexive theses are discussed. Castagnoli observes that many of the notions of entailment assumed by Aristotle presuppose

---

[3]Of course, there are methods by which one may convert an intuitionistic proof of a formula $\varphi \wedge \neg\varphi$ into a proof of $\neg(\varphi \wedge \neg\varphi)$ that are more robust than a mere empty promise. In, say, Gentzen's sequent calculus *LJ* there is obviously an algorithmic method by which when faced with an initial formula $\varphi \wedge \neg\varphi$, one may iteratively apply rules to yield a sequence of formulae terminating with the formula $\neg(\varphi \wedge \neg\varphi)$. What Kapsner's notion of an "empty promise conversion" unearths is the simple fact that possession of such a method is not a *requirement* for the satisfaction of the BHK interpretation of implication.

[4]It is worth noting a similar asymmetry between proving an arbitrary formula from a contradiction and proving a tautology from an arbitrary formula. While the satisfaction of the BHK conditions for, *e.g.*, the formula $\varphi \to (\psi \to \psi)$, is "inert" in the sense that it does not require that a proof of φ be *employed*—the constant function that on any input returns the same proof of $\psi \to \psi$ will suffice, after all—the BHK conditions in this case nevertheless cannot be *vacuously* satisfied. Suggestively, this asymmetry mirrors the choice to emphasize the fact that classical counterexamples to Aristotle's Thesis have inconsistent antecedents rather than the fact that such counterexamples also have tautologous consequents.

that the validity of an inference *demands* that the antecedent play an *active role* in the bringing about of the consequent. The passivity that classical and intuitionistic entailment permits is entirely ruled out in cases in which Aristotle understands entailment as a type of *demonstration*. For example, Aristotle describes a number of necessary conditions for the validity of a demonstration in the *Posterior Analytics*, in which we find:

> The premisses must be the causes of the conclusion, better known than it, and prior to it; its causes, since we possess scientific knowledge of a thing only when we know its cause; prior, in order to be causes; antecedently known, this antecedent knowledge being not our mere understanding of the meaning, but knowledge of the fact as well.[4]

Moreover, Aristotle's interpretation of the notion of demonstration presupposes another feature absent in the case of "empty promise" conversions: Explanation. Castagnoli, for example, points to Robin Smith's commentary on the *Prior Analytics*, in which he considers Aristotle's articulation of connexive theses and suggests that Aristotle's embrace of connexive principles is a natural consequence of his demand for explanation in valid entailments:

> Aristotle... may also have in mind that the conclusion is not *explained* by the premises in such a case [*i.e.*, Aristotle's Thesis]... [O]ne possible reason for seeking a further explanation might be Aristotle's concern, not simply with deductions in general, but with causal or explanatory deductions (demonstrations).[5, p. 190]

It seems reasonable to suggest that, insofar as an inconsistent or logically impossible antecedent cannot *obtain*, such antecedents can never serve as a *cause* of any consequent, including its own negation.[5] This neatly provides an analogy between Wansing and Kapsner's critique of the BHK account of falsification and Aristotle's connexive critique of self-refutation.

3 Contents of the Special Issue

This special issue has assembled a collection of papers both old and new. In addition to papers authored specifically for inclusion in the volume, we have included a number of papers that have been omitted from the mainstream literature on connexive

---

[5]Note that Priest's account in [35], whose account of entailment demands that a contradiction does not entail itself, comports with such an Aristotelian line.

logic, due to being unpublished, untranslated, or otherwise inaccessible. This special issue has assembled a collection of papers both old and new and was projected in connection with the Workshop on Connexive Logic at the Fifth World Congress and School on Universal Logic in Istanbul, June 2015.[6]

We are confident that the scope and breadth of the papers collected in this volume serve to showcase the richness and variety of the contexts in which connexive theses appear. Now, we will proceed to sketch out these themes as we describe the material found in the papers in this volume.

3.1 Posthumous, Translated, and Reprinted Contents

We are especially happy to include a pair of previously unpublished papers that together unearth an important—but forgotten—chapter in the development of connexive logic. These posthumous papers—R. B. Angell's *Connexive Implication, Modal Logic and Subjunctive Conditionals* and Richard Sylvan (formerly "Routley")[7] and Hugh Montgomery's *Models for Connexive Logics*—are especially notable as they jointly reveal the influence that Angell and McCall's "American plan" for connexive logic in [2] and [24] provided the foundation of for the possible-worlds "Australian plan" of Montgomery and Sylvan's [44], Sylvan's [43], Mortensen's [28], and Brady's [7]. At first blush, the matrix semantics favored by McCall and Angell seem entirely distinct from the possible worlds semantics employed in the Australasian papers. Despite this appearance of independence, these two papers document the way in which Angell's semantics for necessity operators in connexive logic were reconfigured and molded into the prototype for the possible worlds approach to connexive logic championed in [44].

Connexive Implication, Modal Logic and Subjunctive Conditionals is the full text of a paper presented at the 1967 meeting of the Association for Symbolic Logic. This work has previously appeared only as the abstract [3]. In the paper, Angell considers formulations of the connexive systems PA1 and CC1 as modal systems—Angell calls their modal reformulations PA1m and CC1m, respectively—in which connexive

---

[6]The website on connexive logic at `https://sites.google.com/site/connexivelogic/`, in addition to offering a very brief overview of connexive logic, collects some information about events on connexive logic and keeps an updated list of publications related to connexive logic.

[7]Due to the depth of the convictions that led Richard Sylvan to take the surname "Sylvan," the matter of how to attribute this paper was an especially sensitive one, requiring a few remarks concerning our decision to attribute co-authorship of the paper to "Richard Routley." Most importantly, although the work—composed while Sylvan was still publishing under the name "Routley"—was neither published nor widely distributed, the paper has on several occasions been cited in print with its authorship attributed to "Routley and Montgomery." We suspect that introducing a competing attribution for the paper would carry a risk of confusion.

implication is not taken as primitive, but instead as a defined connective arising from the logical signature of conjunction, negation, and a necessity operator $\Box$. By observing that the matrices for negation and necessity in PA1m and CC1m are equivalent to those employed in consistency proofs for Lewis' S3, Angell is able to suggest that these connexive logics can be considered to disagree with the more common deductive systems primarily in terms of their interpretation of *conjunction*. Thus, Angell provides a novel perspective in which connexive logic—which is frequently considered to follow from a deviant account of the *conditional*—can just as easily be considered a deviant theory of conjunction.

In the case of Sylvan and Montgomery's *Models for Connexive Logics*, while its *existence* had been known—it had been cited in Mortensen's [28] and in Sylvan's own [43]—its contents have remained mysterious. Serendipitously, the unpublished *Models for Connexive Logic* is primarily an investigation into Angell's own unpublished paper, making the pairing of the papers together all the more appropriate. From a formal perspective, Sylvan and Montgomery's paper complements Angell's paper due to its including proofs of a number of the assertions from *Connexive Implication, Modal Logic and Subjunctive Conditionals* that Angell left unproven. The primary device used in its proofs, too, is unique: Sylvan and Montgomery investigate Angell's PA1m and CC1m by means of a Kripke-style possible worlds semantics. This semantics underscores the relationship between connexive logic and S3 identified by Angell but—more importantly—directly shows how the truth-functional semantics of McCall and Angell gave rise to the possible-worlds semantics employed by Sylvan, Montgomery, Mortensen, and Brady.

We are also pleased to reprint a pair of Grigory Olkhovikov's papers on the connexive logic *LImp* that have been notoriously difficult to access. We include a translation of Olkhovikov's *On a New Three-Valued Paraconsistent Logic* from the Russian [31], in which Olkhovikov introduces the propositional logic *LImp* that counts a number of connexive theses as theorems.[8] Notably, in contrast to the interpretative opaqueness of many multiple-valued semantics for connexive logics, Olkhovikov provides a philosophical discussion of natural language propositions with truth-value gaps and is led on philosophical grounds to the development of semantics corresponding to *LImp*. Complementing this paper is a corrected version of Olkhovikov's extended abstract that had been written for a special session of the Fourth Irish Conference on the Mathematical Foundations of Computer Science and Information Technology in 2006. The abstract was published in the locally produced proceedings volume [32] that was distributed to participants of the conference but does not appear in the more widely-distributed proceedings volume [45]. That this

---

[8]The system *LImp* has been independently discovered by Hitoshi Omori in [33].

pair of papers has only been available by locally-published proceedings volumes has crippled access to the fruits of Olkhovikov's research. We are very pleased to have the opportunity to make these papers easily accessible to scholars investigating connexive logics.

3.2 Contents Original to This Volume

The contributed papers included in this volume provide a showcase of the many points of intersection between connexive principles and a host of independent subfields in philosophical and mathematical logic. In the papers original to the present volume, the understanding of connexive theses is reinforced by examining their intersection with relevant logic, proof-theoretic semantics, conditional logic, traditional Aristotelian theories of syllogism, and Brouwer-Heyting-Kolmogorov semantics, among other areas.

A Comparison of Connexive Logics—the contribution jointly authored by Luis Estrada-González and Elisángela Ramírez-Cámara—revisits the intimate relationship between connexive implication and the traditional syllogism that has surfaced on a number of occasions. This relationship appears, for example, in Hugh MacColl's formalization of traditional syllogistic in [21] and in Storrs McCall's [26], both of which show that connexive theses about entailment lead naturally to the Aristotelian account of the syllogism. Estrada-González's [9] introduced a contraclassical deductive system MRS^P that is "demi-connexive" in the sense that it exhibits some—but not all—of the archetypal connexive properties, *e.g.*, the system counts Aristotle's Theses as theorems while failing to satisfy Boethius' Theses. *A Comparison of Connexive Logics* explicitly analyzes the system MRS^P by the lights of connexive logic. That MRS^P does not validate *all* traditional connexive theses motivates a detailed examination of the constellation of principles that have been given this label and a comparison of the merits of particular properties.

Thomas Ferguson's *On Arithmetic Formulated Connexively* investigates the prospects for formulating mathematics against a connexive background logic. Such investigations have appeared within the connexive logic literature before—McCall's [25] and J. E. Wiredu's [54] consider the impact that connexive principles would have on set or class theories. In particular, the paper considers some philosophical affinities between connexive principles and constructive mathematics. The paper ultimately focuses on three connexive logics—Angell's PA1 and PA2 and Priest's connexive logic—and investigates how weak subsystems of Peano arithmetic fare when analyzed in arbitrary first-order extensions of these systems.

We have observed that the genesis of the field of *conditional logics* was intertwined with connexive principles, although frequently in a veiled fashion. David

Lewis, for example, briefly describes a conditional in [16] that is connexive without identifying it as such. Matthias Unterhuber—whose [48] includes much discussion of the role of connexive theses in conditional logic—sheds further light on this interaction in *Beyond System **P**-Hilbert-Style Convergence Results for Conditional Logics with a Connexive Twist*. Aside from making contributions to conditional logic proper by proving the equivalence of two systems of conditional logic, Unterhuber also strengthens a result that a broad class of conditional logics cannot be enriched with axioms corresponding to Aristotle and Boethius' theses, although they *do* admit *default rule versions* of these axioms. Moreover, the contribution considers the role that such default assumptions play in the broader field of conditional logic and how they lead to the equivalence of a number of monotonicity assumptions with respect to conditional reasoning.

An increasingly frequent approach to connexive logic derives from Heinrich Wansing's "narrow scope" account of negated conditionals described [50], in which the formula $\neg(\varphi \to \psi)$ is identified with the formula $\varphi \to \neg\psi$. This approach can be naturally recalibrated to provide a connexive account of Cecylia Rauszer's *co-implication* connective (see, *e.g.*, [41]). Such a definition directly leads to a connexive counterpart to the bi-intuitionistic logic **2Int** of [53]. Wansing's contribution to this volume—*Natural Deduction for Bi-Connexive Logic and a Two-Sorted Typed λ-Calculus*—provides a deeper analysis of this connexive counterpart—**2C**—by providing a Curry-Howard-style correspondence with a typed λ-calculus 2λ. The ensuing formulae-as-types interpretation of **2C** is discussed and explored, and its utility in giving a deeper analysis of Wansing's systems is described. The formulae-as-types account is especially interesting for the light it sheds on some of the unusual features of Wansing's approach, such as its having inconsistent theorems.

The theme of providing connexive interpretations of Rauszer's co-intuitionistic connectives is again visited in Norihiro Kamide and Heinrich Wansing's joint contribution *Completeness of Connexive Heyting-Brouwer Logic*, in which a number of theorems are proven concerning a connexive version of Heyting-Brouwer logic called BCL (introduced by Wansing in [51] as "I_2C_2"). Both a Gentzen-style sequent calculus and a method of tableau proof are introduced in the paper, notably permitting novel opportunities to relate BCL—and connexive principles more generally—to other systems. The sequent calculus for BCL, for example demands a faithful translation mapping BCL to bi-intuitionistic logic, making new connections apparent. The tableau system considered by the authors borrows the framework of [37] and [38], relating BCL to the class of many-valued modal logics in the style of [38].

Hitoshi Omori's first contribution *A Simple Connexive Extension of the Basic Relevant Logic* **BD** recalls the fragile alliance between connexive and relevance logics witnessed by, *e.g.*, the sections on connexive implication McCall provided for

Anderson and Belnap's [1]. One element of this relationship is revisited in Omori's paper, which applies connexive theses to the investigation of an open problem in relevant logic. In relevant logic, Priest and Sylvan in [39] provides simplified semantics for the relevant logic **BD** in two flavors—a version with the Routley star and a four-valued version without the Routley star—and describes two possible (not connexive) model-theoretic accounts of the falsity conditions for relevant conditionals. Priest and Sylvan leave open the problem of providing a natural proof theory for these particular falsity conditions and, although the problem remains open, Omori's results provide new vantage points from which to frame the problem. Omori introduces three distinct accounts of the falsity conditions for relevant conditionals not considered by Priest and Sylvan—including the aforementioned "narrow scope" account—and proves the correspondence between their semantics and naturally formulated axioms, which leads to new hope for solving the open problem.

In his contribution to the special issue, Nissim Francez identifies two intriguing ways in which negation and conditionals interact in natural language and investigates their properties. The resulting investigation—*Natural-Deduction for Two Connexive Logics*—provides illustrations of these interactions by means of formal dialogues and, by showing a correspondence between certain logical interderivabilities and these particular dialogues, shows the phenomena to be essentially connexive in character. To further study these interactions, Francez introduces a pair of natural deduction calculi $\mathcal{N}^{\neg r}$ and $\mathcal{N}^{\neg l}$—each with a distinct negation operator—in which negation and the conditional exhibit behavior analogous to the natural language, connexive phenomena. Furthermore, Francez' paper introduces connexive logic into the field of proof-theoretic semantics by providing such semantics for the natural deduction systems and showing how the proof-theoretic use of these instances of negation can be understood as meaning-conferring.

An interesting feature of Francez' paper inspired the second contribution by Hitoshi Omori. While Francez' calculus $\mathcal{N}^{\neg r}$ follows the aforementioned "narrow scope" reading of negated conditionals in its identification of $\neg_r(\varphi \to \psi)$ with $\varphi \to \neg_r\psi$, Francez' calculus $\mathcal{N}^{\neg l}$ admits the interderivability of the two formulae:

$$\neg_l(\varphi \to \psi) \dashv\vdash \neg_l\varphi \to \psi$$

where $\neg_l$ is the unary negation operator at the center of $\mathcal{N}^{\neg l}$. The syntactic resemblance between the interpretation of negated conditionals in Francez' $\mathcal{N}^{\neg l}$ and the "narrow scope" reading described above is clear.[9] In Omori's research note on

---

[9]But note that the interderivability validated in $\mathcal{N}^{\neg l}$ is in many ways unusual, even in the context of connexive logic. The above equivalence does not, for example, enjoy the experimental support that Pfeifer has provided for the "narrow scope" equivalence in [34].

Francez' paper, this proximity is explored from a model-theoretic perspective, in which a deductive calculus including the analogous axiom

$$\neg(\varphi \to \psi) \leftrightarrow (\neg\varphi \to \psi)$$

is introduced by means of an axiomatic proof theory and a corresponding possible worlds semantics. The system is yielded by modifying the "narrow scope" falsity conditions in Wansing's connexive logic of [50]; by describing a semantic falsity condition for conditionals that is cognate with Francez' "unusual" equivalence, Omori shows how this leads to a "pseudo-connexive" counterpart of Wansing's favored connexive system.

4 Acknowledgements

There are a number of individuals whose assistance was essential in the production of the present volume.

In the case of the posthumous material, we owe a great deal to those who carefully preserved the respective manuscripts and to those who generously allowed us to include the material in this volume. We are indebted to the family of R. B. Angell for sharing his papers with us and for the permission to reproduce and publish *Connexive Implication, Modal Logic and Subjunctive Conditionals*. Much of Angell's published and unpublished work—including the typescript from which the present paper was drawn—are hosted by his family at the website `http://rbangell.com/`. Likewise, a special debt is owed to Chris Mortensen, for his careful preservation of the manuscript version of *Models for Connexive Logics* and for providing access to the paper. We are also grateful to Richard Sylvan's literary executor, Nick Griffin, and to the family of Hugh Montgomery for generously granting us permission to include this paper posthumously.

We thank Grigory Olkhovikov for his permitting us to include his papers in this special issue. His active involvement during the preparation and correction of his two papers was extraordinarily helpful; the advice and insight he provided while *On a New Three-Valued Paraconsistent Logic* was being translated into English was especially appreciated.

The authors of the contributed papers have given us a further opportunity to demonstrate the vibrant range of directions in which research into connexive logic can be taken and we thank them for the opportunity to include their work in the present volume. During peer review, many elements of the contributed authors' papers were sharpened and refined because of the knowledge and time of anonymous referees who provided their professional analyses. The diligence and thoughtfulness of our many referees were crucial to assembling this volume and we acknowledge the importance of their assistance.

At College Publications, we appreciate Dov Gabbay's entrusting us with stewardship of this issue of the *IFCoLog Journal of Logics and Their Applications*. Furthermore, the editing of this volume has benefited incredibly from the patient and knowledgeable assistance of Jane Spurr.

References

[1] A. R. Anderson and N. D. Belnap, Jr. *Entailment: The Logic of Relevance and Necessity*, volume I. Princeton University Press, Princeton, 1975.

[2] R. B. Angell. A propositional logic with subjunctive conditionals. *Journal of Symbolic Logic*, 27(3):327–343, 1962.

[3] R. B. Angell. Connexive implication, modal logic and subjunctive conditionals. *Journal of Symbolic Logic*, 36(2):367, 1971.

[4] Aristotle. *Posterior Analytics*. Clarendon Press, Oxford, 1928. G. R. G. Mure, translator.

[5] Aristotle. *Prior Analytics*. Hackett, Indianapolis, 1989. R. Smith, editor and translator.

[6] J. L. Bell, D. de Vidi, and G. Solomon. *Logical Options: An Introduction to Classical and Alternative Logics*. Broadview Press, Peterborough, 2001.

[7] R. Brady. A Routley-Meyer affixing style semantics for logics containing Aristotle's thesis. *Studia Logica*, 48(2):235–241, 1989.

[8] L. Castagnoli. *Ancient Self-Refutation: The Logic and History of the Self-Refutation Argument from Democritus to Augustine*. Cambridge University Press, Cambridge, 2010.

[9] L. Estrada-González. Weakened semantics and the traditional square of opposition. *Logica Universalis*, 2(1):155–165, 2008.

[10] T.M. Ferguson. Ramsey's footnote and Priest's connexive logics. *Bulletin of Symbolic Logic*, 20(3):387–388, 2014.

[11] T.M. Ferguson. Logics of nonsense and Parry systems. *Journal of Philosophical Logic*, 44(1):65–80, 2015.

[12] S. Haack. *Deviant Logic: Some Philosophical Issues*. Cambridge University Press, Cambridge, 1975.

[13] L. Humberstone. *The Connectives*. MIT Press, Cambridge, MA, 2011.

[14] A. Kapsner. *Logics and Falsifications: A New Perspective on Constructivist Semantics*. Springer, Cham, 2014.

[15] W. Kneale and M. Kneale. *The Development of Logic*. Oxford University Press, Oxford, 1962.

[16] D. Lewis. *Counterfactuals*. Blackwell, Oxford, 1973.

[17] H. MacColl. The calculus of equivalent statements and integration limits. *Proceedings of the London Mathematical Society*, 9:9–20, 1877.

[18] H. MacColl. The calculus of equivalent statements (II). *Proceedings of the London Mathematical Society*, 9:177–186, 1877.

[19] H. MacColl. The calculus of equivalent statements (III). *Proceedings of the London Mathematical Society*, 10:16–28, 1878.

[20] H. MacColl. The calculus of equivalent statements (IV). *Proceedings of the London Mathematical Society*, 11:113–121, 1879.

[21] H. MacColl. *Symbolic Logic and Its Applications*. Longmans, Green and Company, London, 1906.

[22] H. MacColl. *Hugh MacColl: An Overview of his Logical Work with Anthology*. College Publications, London, 2007. S. Rahman, and J. Redmond, editors.

[23] E. P. Martin and R. K. Meyer. Solution to the *P-W* problem. *Journal of Symbolic Logic*, 47(4):869–887, 1982.

[24] S. McCall. Connexive implication. *Journal of Symbolic Logic*, 31(3):415–433, 1966.

[25] S. McCall. Connexive class logic. *Journal of Symbolic Logic*, 32(1):83–90, 1967.

[26] S. McCall. Connexive implication and the syllogism. *Mind*, 76(303):346–356, 1967.

[27] S. McCall. A history of connexivity. In D. Gabbay et al., editors, *A History of Logic*, volume 11, pages 415–449. Elsevier, Amsterdam, 2012.

[28] C. Mortensen. Aristotle's thesis in consistent and inconsistent logics. *Studia Logica*, 43(1):107–116, 1984.

[29] E. J. Nelson. Intensional relations. *Mind*, 39(156):440–453, 1930.

[30] E. J. Nelson. On three logical principles in intension. *The Monist*, 43(2):268–284, 1933.

[31] G. K. Olkhovikov. On a new three-valued paraconsistent logic. In *Logic of Law and Tolerance*, pages 96–113, Yekaterinburg, 2002. Ural State University Press. In Russian.

[32] G. K. Olkhovikov. Complete, correct and independent axiomatization of first-order fragment of a three-valued paraconsistent logic. In *Proceedings of The Fourth International Conference on Information and The Fourth Irish Conference on the Mathematical Foundations of Computer Science and Information Technology*, pages 245–248, Cork, 2006. University College Cork–National University of Ireland.

[33] H. Omori. From paraconsistent logic to dialetheic logic. In H. Andreas and P. Verdée, editors, *Logical Studies of Paraconsistent Reasoning in Science and Mathematics*. Springer, forthcoming.

[34] N. Pfeifer. Experiments on Aristotle's Thesis: Towards an experimental philosophy of conditionals. *The Monist*, 85(2):223–240, 2012.

[35] G. Priest. Negation as cancellation and connexive logic. *Topoi*, 18(2):141–148, 1999.

[36] G. Priest. *An Introduction to Non-Classical Logic*. Cambridge University Press, Cambridge, first edition, 2001.

[37] G. Priest. *An Introduction to Non-Classical Logic: From If to Is*. Cambridge University Press, Cambridge, second edition, 2008.

[38] G. Priest. Many-valued modal logics: A simple approach. *Review of Symbolic Logic*, 1(2):190–203, 2008.

[39] G. Priest and R. Sylvan. Simplified semantics for basic relevant logics. *Journal of Philosophical Logic*, 21(2):217–232, 1992.

[40] F. P. Ramsey. General propositions and causality. In D. H. Mellor, editor, *Philosophical Papers*, pages 145–163. Cambridge University Press, Cambridge, 1990.

[41] C. Rauszer. A formalization of the propositional calculus of H-B logic. *Studia Logica*, 33(1):23–34, 1974.

[42] S. Read. Hugh MacColl and the algebra of strict implication. *Nordic Journal of Philosophical Logic*, 3(1):59–83, 1998.

[43] R. Routley. Semantics for connexive logics I. *Studia Logica*, 37(4):393–412, 1978.

[44] R. Routley and H. Montgomery. On systems containing Aristotle's thesis. *Journal of Symbolic Logic*, 33(1):82–96, 1968.

[45] A. Seda, M. Boubekeur, T. Hurley, M. Mac an Airchinnigh, M. Schellekens, and G. Strong, editors. *Proceedings of the Irish Conference on the Mathematical Foundations of Computer Science and Information Technology (MFCSIT'06)*, volume 225 of *Electronic Notes in Theoretical Computer Science*. 2009.

[46] R. Sylvan. On reasoning: (ponible) reason for (and also against), and relevance. In D. Hyde and G. Priest, editors, *Sociative Logics and their Applications*, Western Philosophy Series, pages 141–174. Ashgate Publishing, Burlington, VT, 2000.

[47] B. E. R. Thompson. Why is conjunctive simplification invalid? *Notre Dame Journal of Formal Logic*, 32(2):248–254, 1991.

[48] M. Unterhuber. *Possible Worlds Semantics for Indicative and Counterfactual Conditionals? A Formal Philosophical Inquiry into Chellas-Segerberg Semantics*. Ontos Verlag (Logos Series), Frankfurt am Main, 2013.

[49] H. Wansing. *The Logic of Information Structures*, volume 681 of *Springer Lecture Notes in AI*. Springer-Verlag, Berlin, 1993.

[50] H. Wansing. Connexive modal logic. In R. Schimdt, I. Pratt-Hartmann, M. Reynolds, and H. Wansing, editors, *Advances in Modal Logic*, volume 5, pages 367–383. College Publications, London, 2005.

[51] H. Wansing. Constructive negation, implication, and co-implication. *Journal of Applied Non-Classical Logics*, 18(2–3):341–364, 2008.

[52] H. Wansing. Connexive logic. In E. N. Zalta, editor, *The Stanford Encyclopedia of Philosophy*. Fall 2015 edition, 2015.

[53] H. Wansing. Falsification, natural deduction and bi-intuitionistic logic. *Journal of Logic and Computation*, 26(1):425–450, 2016. First published online July 17, 2013.

[54] J. E. Wiredu. A remark on a certain consequence of connexive logic for Zermelo's set theory. *Studia Logica*, 33(2):127–130, 1974.

[55] J. Woods. Two objections to system CC1 of connexive implication. *Dialogue*, 7(3):473–475, 1968.

Received August 2016

Connexive Implication, Modal Logic and Subjunctive Conditionals

Richard Bradshaw Angell
Department of Philosophy
Wayne State University, United States

Abstract

McCall's article on "connexive implication" [4] credits Meredith with the suggestion that the matrices for →, ·, and ∼ in connexive implication can be replaced by the matrices for · and ∼ supplemented by a matrix for the unary necessity operator, □. Since the matrices for McCall's connexive implication are those of this writer's subjunctive conditional [1], it follows that the latter, also, may be reduced to a modal logic.

In this paper two axiom sets for a modal logic based on Meredith's observation are constructed. Both are shown consistent, and they are shown complete with respect to (1) this writer's system, PA1, and (2) McCall's system CC1, respectively. Certain differences between these two systems are pointed out, together with observations concerning Post-completeness and functional completeness in the latter. Finally, a brief discussion is presented concerning some philosophical implications of finding connexive implication, or subjunctive conditionals, reducible to a modal logic.

Professor Storrs McCall and I share an interest in logical systems which contain the non-classical theorems:

1. $\sim(p \to \sim p)$—It is false that if p then not-p,

2. $(p \to q) \to \sim(p \to \sim q)$—If (if p then q) then it is false that (if p then not-q);

although our motives differ. I called my system PA1 (in [1]) a logic of subjunctive conditionals; he called his system, CC1 (in [4]) a system of "connexive implication" and allied himself with those who, according to Sextus Empiricus, "say that a conditional is sound when the contradictory of its consequent is incompatible with its

This paper was supported by NSF Grants GS 630 and GS 1010.

antecedent". Nevertheless, our two systems are very closely related formally. McCall, in effect, added five axioms to my system and established the completeness of this expanded axiom-set with respect to the same truth-tables I had used in 1962 for the primitives, the conditional, conjunction and negation. PA1 showed it possible to have a consistent propositional logic which 1) contains the classical PM [*Principia Mathematica*] calculus (with "⊃" interpreted as "not...or..."), 2) eliminates all the so-called paradoxes of material and strict implication from the conditional, 3) includes most of the traditional logical principles involving conditionals, and yet 4) includes the non-classical theorems mentioned above. McCall pointed out the independence of all such systems of any of the well-known systems of logic and proved his system Post-complete.

Besides these two systems there are many other constructible systems which share the properties just described. The problem is to find a satisfactory one. Certain difficulties of interpretation arose in connection with PA1 which led me to look for better systems; these difficulties are aggravated, rather than mollified, by the new axioms in McCall's expansion, although from a formal point of view his system is certainly the more interesting. These difficulties, as well as the relationships of these two systems to each other and to modal logic, stand out clearly in the light of an observation which McCall credits to Meredith—namely, that the truth-table we both used for the conditional can be eliminated in favor of a unary modal operator.

In this paper I present two modal logics, PA1m and CC1m, which use C.I. Lewis's primitives for possibility, negation and conjunction, and Lewis's definitions of other logical constants, but yield respectively my so-called "logic of subjunctive conditionals" and McCall's system of "connexive implication". The four-valued truth-tables for negation and possibility are those of Lewis's Group II matrices; the truth-table for conjunction is that of PA1 and CC1, not Lewis's. On this basis, the defined conditional comes out to have the same truth-table as that assigned in PA1 and CC1. This suggests the odd conclusion that the difference between PA1 and CC1 on the one hand and Lewis's systems was not related to conditionality or possibility so much as to the different concepts of conjunction.

Table I shows three axiom sets: Lewis's S3, the modal version, PA1m of my logic of subjunctive conditionals and a modal version, CC1m, of McCall's system of "connexive implication". The matrices establish the consistency of the various systems presented, and the derivations appended to this paper show that CC1m and PA1m are complete with respect to McCall's CC1, and my PA1, respectively. *Table II* shows the interrelationships between the axioms and theorems of Lewis's systems, S1, S2, S3, S4, S5, McCall's CC1, my PA1, PA1m, CC1m, and Rosser's axiomatization of the classical propositional calculus of *Principia Mathematica*. The following remarks draw together and reflect upon some of the results shown in these

I *Primitive Symbols*

    1.   Grouping Devices: $(,)$
    2.   Logical Constants: $\sim, \cdot, \Diamond$
    3.   Propositional Variables: p, q, r, s, p'

II *Rules of Formation*

    1.   A single variable, by itself, is well-formed.
    2.   If S is well-formed, then $\ulcorner \sim S \urcorner$ and $\ulcorner \Diamond S \urcorner$ are well-formed.
    3.   If S and S' are well-formed, then $\ulcorner S \cdot S' \urcorner$ is well-formed.

III *Abbreviations (Definitions)*

    1. $\ulcorner (S \vee R) \urcorner$ for $\ulcorner \sim(\sim S \cdot \sim R) \urcorner$
    2. $\ulcorner (S \supset R) \urcorner$ for $\ulcorner \sim(S \cdot \sim R) \urcorner$
    3. $\ulcorner (S \equiv R) \urcorner$ for $\ulcorner (S \supset R) \cdot (R \supset S) \urcorner$
    4. $\ulcorner (S \to R) \urcorner$ for $\ulcorner \sim\Diamond(S \cdot \sim R) \urcorner$
    5. $\ulcorner (S \leftrightarrow R) \urcorner$ for $\ulcorner (S \to R) \cdot (R \to S) \urcorner$
    6. $\ulcorner \Box S \urcorner$ for $\ulcorner \sim\Diamond\sim S \urcorner$

IV *Rules of Transformation (Rules of Inference)*

    1.   If $\vdash S$ and $\vdash (S \to R)$ then $\vdash R$. (*Modus Ponens*)
    2.   If $\vdash S$ and $\vdash R$ then $\vdash (S \cdot R)$ (*Adjunction*)
    3.   If $\vdash S$ and $\vdash S'$ is formed from S by substituting some wff at every occurrence of a proposi-tional variable in S, then $\vdash S'$. (*Substitution*)
    4.   If $\vdash (S \leftrightarrow R)$ and $\vdash Q$, then if Q' is formed replacing an occurrence of S in Q by R, then $\vdash Q'$. (*Rule of Replacement of Strict Equivalents*)[1]

Matrices for Consistency Proofs

$\sim$	p
4	1
3	2
2	3
1	4

$\Diamond$	p
1	1
2	2
1	3
4	4

$(p \cdot q)$	1	2	3	4
1	1	2	3	4
2	2	2	4	4
3	3	4	3	4
4	4	4	4	4

$(p \cdot q)$	1	2	3	4
1	1	2	3	4
2	2	1	4	3
3	3	4	3	4
4	4	3	4	3

(Satisfied in **S3**, **PA1m**, and   (Satisfied in **S3**, not others)   (Satisfied in **PA1m**, **CC1m** only)
    CC1m)

Axiom Systems

S3 Axioms [3]	PA1m Axioms [1]	CC1m Axioms [4]
A1. $(p \cdot q) \to (q \cdot p)$	A1. $p \to p$	A1. $\Box\Diamond(p \cdot p)$
A2. $(q \cdot p) \to p$	A2. $(q \cdot p) \supset p$	A2. $\sim\sim\Diamond(p \cdot p) \to (q \to q)$
A3. $p \to (p \cdot p)$	A3. $p \supset (p \cdot p)$	A3. $p \supset (p \cdot p)$
A4. $(p \cdot (q \cdot r)) \to (q \cdot (p \cdot r))$	A4. $(p \cdot (q \cdot r)) \to (q \cdot (p \cdot r))$	A4. $(p \cdot (q \cdot r)) \to (q \cdot (p \cdot r))$
A5. $((p \to q) \cdot (q \to r)) \to (p \to r)$	A5. $((r \cdot p) \cdot \sim(q \cdot r)) \to (p \cdot \sim q)$	A5. $((r \cdot p) \cdot \sim(q \cdot r)) \to (p \cdot \sim q)$
A6. $\sim\Diamond p \to \sim p$	A6. $\sim\Diamond p \to \sim p$	A6. $\sim\Diamond p \to \sim p$
A7. $(p \to q) \to (\sim\Diamond q \to \sim\Diamond p)$	A7. $(p \to q) \to (\sim\Diamond q \to \sim\Diamond p)$	A7. $(p \to q) \to (\sim\Diamond q \to \sim\Diamond p)$
	A8. $(p \to q) \to \sim(p \to \sim q)$	A8. $p \to ((p \cdot p) \cdot p)$
		A9. $(p \cdot p) \to \Box(p \cdot p)$
		A10. $(p \cdot \sim\Box p) \supset ((q \vee q) \to p)$

Figure 1: A Comparison of **S3**, **PA1m** and **CC1m**

Classical Theorems	S1	S2	S3	S4	S5	PA1	PA1m	CC1	CC1m	PM [2]
1. $p \to p$	+	+	+	+	+	*	A1	*23	*2	+
2. $(p \cdot q) \to (q \cdot p)$	A1	A1	A1	A1	A1	*20	*4	*27	*4	+
3. $(q \cdot p) \to p$	A2	A2	A2	A2	A2	−	−	−	−	+
4. $(q \cdot p) \supset p$	+	+	+	+	+	+	A2	+	*83	+
5. $(p \cdot q) \supset p$	+	+	+	+	+	A8	*73	*92	*84	A2
6. $p \to (p \cdot p)$	A3	A3	A3	A3	A3	−	−	−	−	+
7. $p \supset (p \cdot p)$	+	+	+	+	+	A9	A3	A10	A3	A1
8. $(p \cdot (q \cdot r)) \to (q \cdot (p \cdot r))$	A4	A4	A4	A4	A4	A4	A4	A5	A4	+
9. $((p \to q) \cdot (q \to r)) \to (p \to r)$	A5	A5	A5	A5	A5	+	*66	+	*66	+
10. $(p \to q) \to ((q \to r) \to (p \to r))$	−	−	+	+	+	*43	*44	A1	*44	+
11. $(q \to r) \to ((p \to q) \to (p \to r))$	−	−	+	+	+	A1	*31	*24	*31	+
12. $((r \cdot p) \cdot \sim(q \cdot r)) \to (p \cdot \sim q)$	+	+	+	+	+	+	A5	+	A5	+
13. $(p \supset q) \supset (\sim(q \cdot r) \supset \sim(r \cdot p))$	+	+	+	+	+	*74	*67	*94	*67	A3
14. $(p \to q) \to ((r \cdot p) \to (q \cdot r))$	−	+	+	+	+	A2	*3	+	*3	+
15. $(p \to \sim(q \cdot r)) \to ((q \cdot p) \to \sim r)$	+	+	+	+	+	A3	*59	+	*59	+
16. $((p \cdot q) \to r) \to ((p \cdot \sim r) \to \sim q)$	+	+	+	+	+	+	+	*160	+	+
17. $(p \to \sim q) \to (q \to \sim p)$	+	+	+	+	+	A5	*35	*40	*35	+
18. $\sim\sim p \to p$	+	+	+	+	+	A6	*7	*64	*7	+
19. $(p \to q) \to (p \supset q)$	+	+	+	+	+	A7	*1	*89	*73	+
20. $((p \to p) \to q) \to q$	−	+	+	+	+	−	−	A2	*94	+
21. $(q \cdot q) \to (p \to p)$	+	+	+	+	+	−	−	A4	*96	+
22. $(p \cdot p) \to ((p \to p) \to (p \cdot p))$	−	−	−	−	−	−	−	A6	*112	+
23. $p \to ((p \cdot p) \cdot p)$	+	+	+	+	+	−	−	A7	A8	+
24. $((p \to \sim q) \cdot q) \to \sim p$	+	+	+	+	+	+	*64	A8	*64	+
25. $(p \cdot \sim(p \cdot \sim q)) \to q$	+	+	+	+	+	+	*65	A9	*65	+
26. $(\sim p \vee ((p \to p) \to p)) \vee (((p \to p) \vee (p \to p)) \to p)$	−	−	−	−	−	−	−	A11	*121	+

Non-Classical Theorems	S1	S2	S3	S4	S5	PA1	PA1m	CC1	CC1m	PM
27. $(p \to q) \to \sim(p \to \sim q)$	−	−	−	−	−	A10	A8	*102	+	−
28. $(p \to p) \to \sim(p \to \sim p)$	−	−	−	−	−	+	*2	A12	*108	−
29. $\sim(p \to \sim p)$	−	−	−	−	−	*77	+	+	+	−

Modal Theorems	S1	S2	S3	S4	S5	PA1	PA1m	CC1	CC1m	PM
30. $\sim\Diamond p \to \sim p$	A6	A6	A6	A6	A6		A6		A6	
31. $\Diamond(p \cdot q) \to \Diamond p$	−	A7	+	+	+		−		−	
32. $(p \to q) \to (\sim\Diamond q \to \sim\Diamond p)$	−	−	A7	+	+		A7		A7	
33. $\Diamond\Diamond p \to \Diamond p$	−	−	−	A7	+		−		−	
34. $\Diamond p \to \Box\Diamond p$	−	−	−	−	A7		−		−	
35. $\Box\Diamond(p \cdot p)$	−	−	−	−	−				A1	
36. $\sim\sim\Diamond(q \cdot q) \to (p \to p)$	+	+	+	+	+				A2	
37. $(p \cdot p) \to \Box(p \cdot p)$	−	−	−	−	−				A9	
38. $(p \cdot \sim\Box p) \supset ((q \vee q) \to p)$	−	−	−	−	−				A10	

Figure 2: Interrelationships of Axioms in S1, S2, S3, S4, S5, PA1, PA1m, CC1, CC1m, PM (Rosser)

two tables.

The differences between S3 and PA1m are not so great, in one respect, as they first appear to be. The formulas appearing as $A1$ and $A5$ in each are mutually derivable in the other; Axioms 4, 6, and 7 are identical in both systems. Thus, the real differences boil down to the fact that the strict implications in S3's Axioms 2 and 3 are merely the corresponding truth-functional conditionals in PA1m, and that PA1m contains, in Axiom 8, the non-classical formula, $(p \to q) \to \sim(p \to \sim q)$.

Examination of *Table II* shows that in McCall's and my systems the formulae 3. $(q \cdot p) \to p$ and 6. $p \to (p \cdot p)$ are nowhere derivable. As Everett Nelson pointed [out] long ago, the non-derivability of the first of these, Simplification, is a price we must pay for using the non-classical theorems with standard transposition, syllogism, and the ordinary rules of substitution. Both McCall's system and mine must face up to the demand that we either revise our systems to include these theorems, or explain why they are non-derivable and justify their non-inclusion. In modal logic, the non-derivability of these formulae leads to the non-inclusion of the distinctive axiom of S2, $\Diamond(q \cdot p) \to \Diamond p$. But $\Diamond(q \cdot p) \supset \Diamond p$ also fails in all these systems; and this is clearly due to the properties of conjunction (as reflected in the different conjunction matrix).

Secondly, all of McCall's and my systems include, as intended, the following theorems or axioms which are not derivable in Lewis or in classical logic:

27. $(p \to q) \to \sim(p \to \sim q)$
28. $(p \to p) \to \sim(p \to \sim p)$
29. $\sim(p \to \sim p)$

and others. Ordinarily, this would be cause for self-congratulation. But when these theorems are reduced to modal propositions they present serious problems of interpretation. They become, respectively, equivalent to theorems stating:

---

[1]McCall [4] proves that this rule is derivable in CC1, using theorems which can be established in S3, PA1 and CC1, though not in S1 or S2. Hence $R4$ is derivable in all three systems.

[2]Notes on Table II:

- "+" means that the theorem to the left is derivable in the system indicated above it.

- "−" means that the theorem to the left is provably not derivable in this system. S1, S2, S3, S4, S5 are based on the formulations in [2], except that "$p \to \Diamond p$" and "$(p \to q) \to (\Diamond p \to \Diamond q)$" are replaced by the axioms "$\sim \Diamond p \to \sim p$" and "$(p \to q) \to (\sim \Diamond q \to \sim \Diamond p)$" of [3].

- PA1 refers to the system of [1].

- CC1 refers to the system of connexive implication in [4].

- PA1m and CC1m are the modal versions of PA1 and CC1.

- "$*n$" gives the number of the theorem in this system indicated at the top of the column.

27'. $\Diamond(p \cdot q) \to \Diamond(p \cdot \sim q)$
28'. $\Diamond(p \cdot p) \to \Diamond(p \cdot \sim p)$
29'. $\Diamond(p \cdot p)$

The first would seem false whenever q is a tautology; the second would seem false whenever p is consistent; and the third would be false whenever p was inconsistent. These consequences alone seem fairly devastating for both of our systems.

When we consider the new axioms McCall added to establish Post-completeness, however, the difficulties in interpretation increase. To be sure some of additional axioms in CC1 (*cf.* 20, 21, 22, 23, and 26 in Table II) seem plausible, *e.g.*, his CC1 Axiom 2, $(((p \to p) \to q) \to q)$ (#20 in Table II) which occurs in all Lewis systems except S1, as well as in PM. But there is a peculiar and irrational bias in some of them. Thus Axiom 7 of CC1 (#23 in Table II), $p \to ((p \cdot p) \cdot p)$, seems eminently plausible until it is realized that while p will imply (or be implied by) any conjunction containing just an odd number of iterations of itself, it never implies (or is implied by) a conjunction with an even number of conjuncts of itself. Thus

- $p \to ((p \cdot p) \cdot p)$

- $p \to ((((p \cdot p) \cdot p) \cdot p) \cdot p)$

- $p \to ((((((p \cdot p) \cdot p) \cdot p) \cdot p) \cdot p) \cdot p)$

are all logical truths, but

- $p \to (p \cdot p)$

- $p \to (((p \cdot p) \cdot p) \cdot p)$

- $p \to (((((p \cdot p) \cdot p) \cdot p) \cdot p) \cdot p)$

are merely contingent. Again, any conjunction containing just an even number of conjuncts of p, as in

- $(p \cdot p) \to (q \to q)$ (CC1 Axiom 4, #21 in Table II)

- $((p \cdot p) \cdot (p \cdot p)) \to (q \to q)$, *etc.*

will imply any theorem of CC1, but no conjunctions having just an odd number of occurrences of a given variable will imply any theorem. Similar remarks pertain to the double p's in CC1 Axiom 7 (#22 in Table II). It is hard to [see] how the concept of connexive implication—that logically true conditionals have antecedents which are compatible with the contradictories of their consequents—can justify these distinctions between odd and even numbers of variable occurrences. In the modal versions of **CC1**, the modal correlates of these implausibilities (*cf.* 35, 36, 37, 38 in Table II, which represent the axioms (except Axiom 8) of **CC1m** which differ from those of **PA1m**) seem just as unlikely:

35.	$\Box\Diamond(p \cdot p)$	(Axiom 1, **CC1m**)
36.	$\sim\sim\Diamond(p \cdot p) \to (q \to q)$	(Axiom 2, **CC1m**)
37	$(p \cdot p) \to \Box(p \cdot p)$	(Axiom 9, **CC1m**)
38	$(p \cdot \sim\Box p) \supset ((q \vee q) \to p)$	(Axiom 10, **CC1m**)

The only one of these four that is included in the modal systems of Lewis is the second, #36, and this only because it is a paradox of strict implication. The other three are not derivable in any of the five Lewis systems, and in any case are intuitively unconvincing. The peculiarities of a type of conjunction which yields different implications for odd-numbered conjunctive iterations of a variable than for even-numbered ones stand out in all four of these; each fails if an even-numbered conjunctive iteration (or alternation) is replaced by an odd-numbered one. These same peculiarities are reflected in the difficulty of finding a consistent interpretation for the conjunction matrix axiomatized in **CC1**, **PA1**, *etc.*

Among the interesting formal results in McCall's system is that fact that not only can we define connexive implication in terms of negation, conjunction and a modal operator (possibility or necessity), but we can define the modal operators (either possibility or necessity) in terms of the primitives of **CC1**, *i.e.*, negation, conjunction and the conditional. Thus we could have, in **CC1**, the definitions:

- $\Box p =_{Df} ((p \to p) \to p)$

- $\Diamond p =_{Df} \sim((\sim p \to \sim p) \to \sim p)$

Since McCall proved that all tautologies are theorems in his system, and the matrices for those defined terms are identical with those already referred to (the Group II Lewis matrices), it follows that the nine axioms of **CC1m** can also be used for **CC1** with the conditional primitive instead of the modal operator. Thus **CC1** and **CC1m** are exactly equivalent systems. Since **CC1** was proved Post-complete, **CC1m** can be proved Post-complete also. Since **CC1** is functionally incomplete, **CC1m** is functionally incomplete also.

In spite of the instructive and interesting formal properties revealed in **CC1** and **PA1**, in my opinion, the foregoing analysis shows rather conclusively the inadequacy as a formalization of a viable logic, of both my system **PA1** and McCall's **CC1**. Admitting these inadequacies does not, of course, entail rejection of the non-classical theorems. There are other systems which contain these theorems and lack the objectionable features just discussed. Although not fully satisfactory systems, **PA1** and **CC1** are, I believe, helpful first efforts towards the construction of a satisfactory non-classical logic.

1.	$[A6 = 1]$	$(p \to q) \to \sim(p \cdot \sim q)$	PA1 Axiom 7
2.	$[A8 = 2]$	$(p \to p) \to \sim(p \to \sim p)$	CC1 Axiom 12
3.	$[A7 = (A5 \to 3)]$	$(p \to q) \to ((r \cdot p) \to (q \cdot r))$	PA1 Axiom 2
4.	$[3 = (A1 \to 4)]$	$(p \cdot q) \to (q \cdot p)$	S3 Axiom 1
5.	$[A7 = (4 \to 5)]$	$\sim\Diamond(q \cdot p) \to \sim\Diamond(p \cdot q)$	
6.	$[D4, 5 = 6]$	$(\sim q \to p) \to (\sim p \to q)$	
7.	$[6 = (A1 \to 7)]$	$\sim\sim p \to p$	PA1 Axiom 6^3
8.	$[3 = (7 \to 8)]$	$(q \cdot \sim\sim p) \to (p \cdot q)$	
9.	$[A7 = (8 \to 9)]$	$(p \to q) \to (\sim q \to \sim p)$	
10.	$[9 = (4 \to 10)]$	$\sim(q \cdot p) \to \sim(p \cdot q)$	CC1 Axiom 9
11.	$[D4, 5 = (A1 \to 11)]$	$\sim\Diamond(\sim p \cdot p)$	
12.	$[A7 = (8 \to 12)]$	$\sim\Diamond(\sim p \cdot p) \to \sim\Diamond(p \cdot \sim\sim\sim p)$	
13.	$[12 = (11 \to 13)]$	$p \to \sim\sim p$	
14.	$[3 = (4 \to 14)]$	$(\sim(q \cdot r) \cdot (p \cdot r)) \to ((r \cdot p) \cdot \sim(q \cdot r))$	
15.	$[A7 = (14 \to 15)]$	$\sim\Diamond((r \cdot p) \cdot \sim(q \cdot r)) \to \sim\Diamond((q \cdot r) \cdot (p \cdot r))$	
16.	$[9 = (15 \to 16)]$	$\sim\sim\Diamond(\sim(q \cdot r) \cdot (p \cdot r)) \to \sim\sim\Diamond((r \cdot p) \cdot \sim(q \cdot r))$	
17.	$[3 = (16 \to 17)]$	$((p \to q) \cdot \sim\sim\Diamond(\sim(q \cdot r) \cdot (p \cdot r))) \to$ $\qquad (\sim\sim\Diamond((r \cdot p) \cdot \sim(q \cdot r)) \cdot (p \to q))$	
18.	$[A7 = (17 \to 18)]$	$\sim\Diamond(\sim\sim\Diamond((r \cdot p) \cdot \sim(q \cdot r)) \cdot (p \to q)) \to$ $\qquad \sim\Diamond((p \to q) \cdot \sim\sim\Diamond(\sim(q \cdot r) \cdot (p \cdot r)))$	
19.	$[5 = (3 \to 19)]$	$\sim\Diamond(\sim\sim\Diamond((r \cdot p) \cdot \sim(q \cdot r)) \cdot (p \to q))$	
20.	$[18 = (19 \to 20)]$	$\sim\Diamond((p \to q) \cdot \sim\sim\Diamond(\sim(q \cdot r) \cdot (p \cdot r)))$	
21.	$[5 = (20 \to 21)]$	$\sim\Diamond(\sim\sim\Diamond(\sim(q \cdot r) \cdot (p \cdot r)) \cdot (p \to q))$	
22.	$[9 = (5 \to 22)]$	$\sim\sim\Diamond(q \cdot p) \to \sim\sim\Diamond(p \cdot q)$	
23.	$[3 = (22 \to 23)]$	$(r \cdot \sim\sim\Diamond(q \cdot p)) \to (\sim\sim\Diamond(p \cdot q) \cdot r)$	
24.	$[A7 = (23 \to 24)]$	$\sim\Diamond(\sim\sim\Diamond(p \cdot q) \cdot r) \to \sim\Diamond(r \cdot \sim\sim\Diamond(q \cdot p))$	
25.	$[D4, 22 = (21 \to 25)]$	$(p \to q) \to ((p \cdot r) \to (q \cdot r))$	
26.	$[9 = (A7 \to 26)]$	$\sim(\sim\Diamond q \to \sim\Diamond p) \to \sim(p \to q)$	
27.	$[3 = (26 \to 27)]$	$(r \cdot \sim(\sim\Diamond q \to \sim\Diamond p)) \to (\sim(p \to q) \cdot r)$	
28.	$[A7 = (27 \to 28)]$	$\sim(\sim(p \to q) \cdot r) \to \sim\Diamond(r \cdot \sim(\sim\Diamond q \to \sim\Diamond p))$	
29.	$[D4, 5 = (25 \to 29)]$	$\sim\Diamond(\sim((p \cdot r) \to (q \cdot r)) \cdot (p \to q))$	
30.	$[28 = (29 \to 30)]$	$\sim\Diamond((p \to q) \cdot \sim(\sim\Diamond(q \cdot r) \to \sim\Diamond(p \cdot r))$	
31.	$[D4, 30 = 31]$	$(p \to q) \to ((q \to r) \to (p \to r))$	CC1 Axiom 1

32. $[31 = (13 \to 32)]$ $\quad (\sim\sim p \to q) \to (p \to q)$
33. $[31 = 6 \to (32 \to 33)]$ $\quad (\sim q \to \sim p) \to (p \to q)$
34. $[31 = (7 \to 34)]$ $\quad (p \to q) \to (\sim\sim p \to q)$
35. $[31 = 34 \to (33 \to 35)]$ $\quad (p \to \sim q) \to (q \to \sim p)$ $\qquad$ PA1 Axiom 5
36. $[31 = 36]$ $\quad (\sim q \to \sim p) \to ((\sim p \to \sim r) \to (\sim q \to \sim r))$
37. $[31 = 9 \to (36 \to 37)]$ $\quad (p \to q) \to ((\sim p \to \sim r) \to (\sim q \to \sim r))$
38. $[31 = (9 \to 38)]$ $\quad ((\sim p \to \sim r) \to (\sim q \to \sim r)) \to$
$$((r \to p) \to (\sim q \to \sim r))$$
39. $[31 = 37 \to (38 \to 39)]$ $\quad (p \to q) \to ((r \to p) \to (\sim q \to \sim r))$
40. $[31 = 9 \to (39 \to 40)]$ $\quad (p \to q) \to (\sim(\sim q \to \sim r) \to \sim(r \to p))$
41. $[9 = (33 \to 41)]$ $\quad \sim(r \to q) \to \sim(\sim q \to \sim r)$
42. $[31 = (41 \to 42)]$ $\quad (\sim(\sim q \to \sim r) \to \sim(r \to p)) \to$
$$(\sim(r \to q) \to \sim(r \to p))$$
43. $[31 = 42 \to (40 \to 43)]$ $\quad (p \to q) \to (\sim(r \to q) \to \sim(r \to p))$
44. $[31 = 41 \to (33 \to 44)]$ $\quad (p \to q) \to ((r \to p) \to (r \to q))$ $\qquad$ PA1 Axiom 1
45. $[44 = (4 \to 45)]$ $\quad ((r \cdot p) \to (q \cdot r)) \to ((r \cdot p) \to (r \cdot q))$
46. $[44 = 45 \to (3 \to 46)]$ $\quad (p \to q) \to ((r \cdot p) \to (r \cdot q))$
47. $[31 = (4 \to 47)]$ $\quad ((r \cdot p) \to (r \cdot q)) \to ((p \cdot r) \to (r \cdot q))$
48. $[31 = 46 \to (47 \to 48)]$ $\quad (p \to q) \to ((p \cdot r) \to (r \cdot q))$ $\qquad$ CC1 Axiom 3
49. $[46 = (7 \to 49)]$ $\quad ((q \cdot p) \cdot \sim\sim r) \to ((q \cdot p) \cdot r)$
50. $[31 = 49 \to (4 \to 50)]$ $\quad ((q \cdot p) \cdot \sim\sim r) \to (r \cdot (q \cdot p))$
51. $[31 = 50 \to (A4 \to 51)]$ $\quad ((q \cdot p) \cdot \sim\sim r) \to (q \cdot (r \cdot p))$
52. $[46 = (4 \to 52)]$ $\quad (q \cdot (r \cdot p)) \to (q \cdot (p \cdot r))$
53. $[31 = 52 \to (A4 \to 53)]$ $\quad (q \cdot (r \cdot p)) \to (p \cdot (q \cdot r))$
54. $[31 = 51 \to (53 \to 54)]$ $\quad ((q \cdot p) \cdot \sim\sim r) \to (p \cdot (q \cdot r))$
55. $[44 = (13 \to 55)]$ $\quad (p \cdot (q \cdot r)) \to (p \cdot \sim\sim(q \cdot r))$
56. $[31 = 54 \to (55 \to 56)]$ $\quad ((q \cdot p) \cdot \sim\sim r) \to (p \cdot \sim\sim(q \cdot r))$
57. $[A7 = (56 \to 57)]$ $\quad (p \to \sim(q \cdot r)) \to ((q \cdot p) \to \sim r)$ $\qquad$ PA1 Axiom 3
58. $[44 = (7 \to 58)]$ $\quad (p \to \sim\sim q) \to (p \to q)$
59. $[31 = 57 \to (58 \to 59)]$ $\quad (p \to \sim(q \cdot \sim r)) \to ((q \cdot p) \to r)$
60. $[31 = (4 \to 60)]$ $\quad ((q \cdot p) \to r) \to ((p \cdot q) \to r)$
61. $[31 = 59 \to (60 \to 61)]$ $\quad (p \to \sim(q \cdot \sim r)) \to ((p \cdot q) \to r)$
62. $[44 = (A6 \to 62)]$ $\quad (p \to \sim\Diamond(q \cdot \sim r)) \to (p \to \sim(q \cdot \sim r))$
63. $[31 = 62 \to (61 \to 63)]$ $\quad (p \to (q \to r)) \to ((p \cdot q) \to r)$ $\qquad$ Importation
64. $[63 = (35 \to 64)]$ $\quad ((p \to \sim q) \cdot q) \to \sim p$ $\qquad$ CC1 Axiom 8
65. $[59 = (A1 \to 65)]$ $\quad (p \cdot \sim(p \cdot \sim q)) \to q$ $\qquad$ CC1 Axiom 9
66. $[63 = (31 \to 66)]$ $\quad ((p \to q) \cdot (q \to r)) \to (p \to r)$ $\qquad$ S3 Axiom 5
67. $[44 = 8 \to (A5 \to 67)]$ $\quad (\sim(q \cdot r) \cdot \sim\sim(r \cdot p)) \to (p \cdot \sim q)$
68. $[9 = (67 \to 68)]$ $\quad (p \supset q) \to (\sim(q \cdot r) \supset \sim(r \cdot p))$
69. $[A6 = (68 \to 69)]$ $\quad (p \supset q) \supset (\sim(q \cdot r) \supset \sim(r \cdot p))$ $\qquad$ PM Axiom 3
70. $[31 = 53 \to (4 \to 70)]$ $\quad (p \cdot (q \cdot r)) \underset{305}{\to} ((p \cdot q) \cdot r)$

71. $[25 = (4 \to 71)]$ $\quad ((p \cdot q) \cdot \sim p) \to ((q \cdot p) \cdot \sim p)$
72. $[9 = (71 \to 72)]$ $\quad \sim((q \cdot p) \cdot \sim p) \to ((p \cdot q) \cdot \sim p)$
73. $[72 = (A2 \to 73)]$ $\quad \sim((p \cdot q) \cdot \sim p)$ $\hfill$ PM Axiom 2

The theorems of PA1m, and their justifications, may be kept provided we replace Theorems 1 and 2 by

1. $[A6 = (A1 \to 1)]$ $\quad \sim\sim\Diamond(q \cdot q)$
2. $[A2 = (1 \to 2)]$ $\quad p \to p$

changing '$A1$' to '2' in the proofs of Theorems 4, 7, and 11, and replacing Theorem 73 in PA1m by what was 2 in PA1m, *i.e.*,

73. $[A6 = 73]$ $\quad (p \to q) \to \sim(p \cdot \sim q)$

The proof of Theorem 2 in PA1m does not hold, so far, in CC1m since '$(p \to q) \to \sim(p \to \sim q)$' must be derived from CC1m Axiom 12, $(p \to p) \to \sim(p \to \sim p)$, which is derivable in proofs in CC1m then may proceed as follows, establishing PA1m's Axiom 2, '$(q \cdot p) \supset p$', in Theorem 83.

74. $[44 = 46 \to (33 \to 74)]$ $\quad (p \to q) \to ((r \cdot \sim q) \to (r \cdot \sim p))$
75. $[31 = 74 \to (9 \to 75)]$ $\quad (p \to q) \to ((r \supset q) \to (r \supset p))$
76'. $[35 = (A6 \to 76')]$ $\quad (q \cdot q) \to \sim\sim\Diamond(q \cdot q)$
76. $[31 = 76' \to (A2 \to 76)]$ $\quad (q \cdot q) \to (p \to p)$
77. $[31 = 76 \to (73 \to 77)]$ $\quad (q \cdot q) \to \sim(p \cdot \sim p)$
78. $[61 = (77 \to 78)]$ $\quad ((q \cdot q) \cdot p) \to p$
79. $[25 = (4 \to 79)]$ $\quad ((r \cdot p) \cdot \sim(q \cdot r)) \to ((p \cdot r) \cdot \sim(q \cdot r))$
80. $[31 = A5 \to (79 \to 80)]$ $\quad ((p \cdot r) \cdot \sim(q \cdot r)) \to (p \cdot r)$
81. $[9 = (80 \to 81)]$ $\quad (p \supset q) \to ((p \cdot r) \supset (q \cdot r))$
82. $[81 = (A3 \to 82)]$ $\quad (q \cdot p) \supset ((q \cdot q) \cdot p)$
83. $[75 = 78 \to (82 \to 83)]$ $\quad (q \cdot p) \supset p$
84. $[72 = (83 \to 84)]$ $\quad (p \cdot q) \supset p$
85. $[51 = (10 \to 85)]$ $\quad (q \cdot \sim(p \cdot q)) \to \sim p$
86. $[A7 = (85 \to 86)]$ $\quad \Box p \to (q \to (p \cdot q))$
87. $[31 = A6 \to (7 \to 87)]$ $\quad \Box p \to p$
88. $[31 = A7 \to (33 \to 88)]$ $\quad (p \to q) \to (\Box p \to \Box q)$
89. $[31 = 13 \to (A2 \to 89)]$ $\quad \Diamond(q \cdot q) \to (p \to p)$
90. $[88 = (89 \to 90)]$ $\quad \Box\Diamond(q \cdot q) \to \Box(p \to p)$

91. $[90 = (A1 \to 91)]$ $\Box(p \to p)$

92. $[86 = (91 \to 92)]$ $q \to ((p \to p) \cdot q)$

93. $[6 = (92 \to 93)]$ $((p \to p) \supset q) \to q$

94. $[31 = 73 \to (93 \to 94)]$ $((p \to p) \to q) \to q$ CC1 Axiom 2

95. $[33 = (A6 \to 95)]$ $p \to \Diamond p$

96. $[31 = 95 \to (89 \to 96)]$ $(q \cdot q) \to (p \to p)$ CC1 Axiom 4

97. $[31 = 9 \to (7 \to 97)]$ $(\sim r \to (p \cdot \sim q)) \to ((p \supset q) \to r)$

98. $[31 = (75 \to 98)]$ $((p \supset q) \to r) \to ((p \to q) \to r)$

99. $[31 = 86 \to (97 \to 99)]$ $\Box p \to ((p \supset q) \to q)$

100. $[31 = 99 \to (98 \to 100)]$ $\Box p \to ((p \to q) \to q)$

101. $[100 = (A1 \to 101)]$ $(\Diamond(p \cdot p) \to \Diamond(p \cdot p)) \to \Diamond(p \cdot p)$

102. $[31 = A7 \to (33 \to 102)]$ $(p \to q) \to (\Diamond p \to \Diamond q)$

103. $[31 = 3 \to (102 \to 103)]$ $(p \to p) \to (\Diamond(p \cdot p) \to \Diamond(p \cdot p))$

104. $[31 = 103 \to (101 \to 104)]$ $(p \to p) \to (p \cdot p)$

105. $[31 = 11 \to (48 \to 105)]$ $(p \cdot q) \to (p \cdot \sim\sim q)$

106. $[102 = (105 \to 106)]$ $\Diamond(p \cdot q) \to \Diamond(p \cdot \sim\sim q)$

107. $[31 = 104 \to (106 \to 107)]$ $(p \to p) \to \Diamond(p \cdot \sim\sim p)$

108. $[31 = 107 \to (13 \to 108)]$ $(p \to p) \to \sim(p \to \sim p)$ CC1 Axiom 12

109. $[100 = 109]$ $\Box(p \cdot p) \to (((p \cdot p) \to (p \cdot p)) \to (p \cdot p))$

110. $[31 = A9 \to (109 \to 110)]$ $(p \cdot p) \to (((p \cdot p) \to (p \cdot p)) \to (p \cdot p))$

111. $[31 = (3 \to 111)]$ $(((p \cdot p) \to (p \cdot p)) \to (p \cdot p)) \to ((p \to p) \to (p \cdot p))$

112. $[31 = 110 \to (111 \to 112)]$ $(p \cdot p) \to ((p \to p) \to (p \cdot p))$ CC1 Axiom 6

113. $[9 = (100 \to 113)]$ $\sim((p \to p) \to p) \to \sim\Box p$

114. $[48 = (113 \to 114)]$ $(\sim((p \to p) \to p) \cdot p) \to (p \cdot \sim\Box p)$

115. $[48 = (7 \to 115)]$ $(\sim\sim p \cdot \sim((p \to p) \to p)) \to (\sim((p \to p) \to p) \cdot p)$

116. $[31 = 115 \to (114 \to 116)]$ $(\sim\sim p \cdot \sim((p \to p) \to p)) \to (p \cdot \sim\Box p)$

117. $[31 = 34 \to (25 \to 117)]$ $(p \to q) \to ((\sim\sim p \cdot \sim r) \to (q \cdot \sim r))$

118. $[31 = 117 \to (9 \to 118)]$ $(p \to q) \to (\sim(q \cdot \sim r) \to \sim(\sim\sim p \cdot \sim r))$

119. $[118 = 116 \to (A10 \to 119)]$ $\sim(\sim\sim(\sim\sim p \cdot \sim((p \to p) \to p)) \cdot \sim((q \vee q) \to p))$

120. $[119 = 120]$ $(\sim p \vee ((p \to p) \to p)) \vee ((q \vee q) \to p)$

121. $[120 = 121]$ $(\sim p \vee ((p \to p) \to p)) \vee$ CC1 Axiom 11
$$(((p \to p) \vee (p \to p)) \to p)$$

References

[1] R. B. Angell. A propositional logic with subjunctive conditionals. *Journal of Symbolic Logic*, 27(3):327–343, 1962.

[2] R. Feys. *Modal Logics*. E. Nauwelaerts, Louvain, 1965.

---

[3] Also in S3 but shown derivable by McKinsey in [5].

[3] C.I. Lewis and C.H. Langford. *Symbolic Logic*. Century, London, 1932.

[4] S. McCall. Connexive implication. *Journal of Symbolic Logic*, 31(3):415–433, 1966.

[5] J.C.C. McKinsey. A reduction in number of the postulates for C.I. Lewis' system of strict implication. *Bulletin of the American Mathematical Society*, 40(6):425–427, 1934.

Received posthumously

MODELS FOR CONNEXIVE LOGICS

RICHARD ROUTLEY
Australian National University, Australia

HUGH MONTGOMERY
University of Auckland, New Zealand

Sentential connexive logics have a two-valued worlds semantics (by [6]). But at least for some connexive logics it is possible to improve vastly on the cumbersome uniform models so supplied. The basic idea for the more improved models here presented is that the worlds of each connexive logic model divide into two classes, the regular worlds U where conjunctions are evaluated in the normal way, and the irregular equivalential worlds I where conjunctions are evaluated like equivalences. That is, the initial evaluation rule for conjunction is as follows:

- for $H \in I$, $v(A \cdot B, H) = T$ iff $v(A, H) = v(B, H)$; and

- for $H \in U = K - I$, $v(A \cdot B, H) = T$ iff $v(A, H) = T = v(B, H)$.

Here A and B are arbitrary wff; K is the set of worlds of the given model, with U and J subsets of K; and v is a two-valued valuation function; (for background see Kripke [3]).

The fact that $\cdot$ ceases to behave normally in I situations is forced by the very features that characterise connexive logics, the inclusion of such connexive principles as *Boethius* $p \to q \to \sim(p \to \sim q)$ and its consequence *Aristotle* $\sim(p \to \sim p)$.[1] This becomes plain when one considers modal reformulations of these principles: *Aristotle* for example can be reformulated as $\Diamond(p \cdot p)$ given quite weak principles generally admitted in connexive logics. If however the modal connective $\Diamond$ is interpreted in anything approaching the usual fashion, as saying that $\Diamond A$ holds iff A holds in some possible situation, then $p \cdot p$ must hold in some situation, whatever p. Were conjunction evaluated normally (*i.e.* as for U worlds) we should have somehow to guarantee that each sentential parameter p and each wff held in some world; and this is not only difficult to ensure but would do too much. For it would make $\Diamond p$, and so $\sim\Box p$, theorems, and hence render connexive logics, all of which contain theses of

---

[1]On the rationale of the names these principles are given, see [7].

the form $\Box B$, inconsistent. Indeed the fact that $\Diamond(p \cdot p)$ is a theorem but $\Diamond p$ is not, on pain of inconsistency, means that within the general framework of a two-valued worlds semantics one has no option but to change the conjunction rule. For were conjunction normal then in every world $p \cdot p$ would hold where and only where p held, so $p \cdot p$ and p would be interchangeable in every sentence context, and hence in the frame $\Diamond(\ )$. The only question then is how conjunction behaves in irregular worlds. There are, as we shall see, several options other than material equivalence that can be exploited in order to distinguish connexive systems.

The semantical features of connexive logics, the fact that such irregular worlds have to be introduced into the modellings, is enough to cast serious doubt at least on the adequacy of such systems as providing analyses of implication and subjunctive conditionals (*cf.* also [7]). Perhaps the observed semantical peculiarities of connexive logics will do something to dampen the philosophical revival these logics are currently enjoying.

1 McCall's system CC1

The most investigated and best understood connexive logic is McCall's system CC1. It makes a natural starting point before we turn to the more difficult but superior systems Angell has devised.

CC1 has primitive connectives the set $\{\sim, \Box, \cdot\}$. The connexive implication $\rightarrow$ is defined:

- $A \rightarrow B =_{Df} \Box \sim (A \cdot \sim B)$.

Alternatively, if $\rightarrow$ is taken as primitive in place of $\Box$, then $\Box$ can be defined:

- $\Box A =_{Df} (A \rightarrow A) \rightarrow A$; or

- $\Box A =_{Df} \mathbf{1} \rightarrow A$ where $\mathbf{1} =_{Df} \sim(p \cdot \sim p)$.

The complex postulate theory of CC1 is developed in McCall [5]. The semantics of CC1, to which we turn, resembles that of finite-valued modal logics, *e.g.* that of Łukasiewicz's system Ł as modelled in Lemmon [4]. A CC1-*model* M is a structure $M = \langle G, J, R, v \rangle$, *i.e.* a two-world structure with $K = \{G, J\}$, where R is reflexive and such that GRJ but not JRG. The valuation function v, specified in the model for atomic wff only is extended inductively to all wff recursively, as follows:

- $v(\sim A, H) = T$ iff $v(A, H) = F$

310

- $v(\Box A, H) = T$ iff $(H_i)(HRH_i \supset v(A, H_i) = T)$

- $v(A \cdot B, H) = T$ iff $v(A, H) = v(B, G)$ and either $H = J$ or $v(A, H) = T$

in each case for every $H \in K$ and every wff. A wff B is CC1-*valid* iff $v(B, G) = T$ in all CC1-models. CC1-satisfiability, *etc.*, are defined in the usual way.

CC1 models can be reformulated in ways that look initially to be more general but turn out not to be; for example G can be replaced by a set U of elements to which G belongs given that it required for H in $K = U \cup \{J\}$ GRH iff $H = J$. Likewise the element J can be replaced by a set of elements I subject to appropriate restrictions.

Theorem 1 (Soundness theorem for CC1). *If* $\vdash_{CC1} B$, *then* B *is* CC1-*valid.*

Proof. By induction over the length of proofs. Alternatively, and more briefly given [5], the result will follow upon showing that every CC1-model provides a CC1-matrix. $\qed$

In the style of Carnap, and later Kripke [3], we represent propositions as mappings from elements of K to $\Pi = \{T, F\}$. In the case of CC1-models there are only four propositions to consider: a first, 1, which maps $\{G, J\}$ to T, 2, which maps G to T (and J to F), 3, which maps J to T, and 4, which maps both G and J to F and accordingly maps $\varnothing$ to T.

A proposition is *designated* if it maps G to T; thus 1 and 2 are designated. Prescriptions for compound mappings $\sim\rho$, $\Box\rho$ and $\rho \cdot \sigma$ in terms of mappings ρ and σ parallel the recursive specification of v. Thus $\sim\rho$ is defined that

- $\sim\rho(H) = T$ iff $\rho(H) = F$,

i.e., $\sim\rho$ maps H to T iff ρ maps H to F. Hence $(\sim\rho)(H) = \sim\rho(H)$.

- $\Box\rho(H) = T$ iff $\rho(H_i) = T$ for every H_i such that HRH_i.

- $\rho \cdot \sigma(H) = T$ iff $\rho(H) = \sigma(H)$ and either $\rho(H) = T$ or $H = J$.

The effects of $\sim$, $\Box$, and $\cdot$ on propositions 1, 2, 3, and 4 can be represented in the following matrices S:

$\cdot$	1	2	3	4		$\sim$			$\Box$	
$\star 1$	1	2	3	4		$\star 1$	4		$\star 1$	1
$\star 2$	2	1	4	3		$\star 2$	3		$\star 2$	4
3	3	4	3	4		3	2		3	3
4	4	3	4	3		4	1		4	4

Lemma 1. *If B is CC1-valid then B takes a designated value for each assignment of propositions to its atomic parts, and conversely.*

Lemma 2. *The matrices (S) are characteristic for CC1.*

Proof. The result is proved in McCall [5]. □

Theorem 2 (Completeness theorem for CC1). *If B is CC1-valid then $\vdash_{CC1} B$.*

Proof. By the previous lemmata. □

2 Angell's systems

As an analysis of implication CC1 is a remarkably unsatisfactory system. For example it contains as axioms the paradoxical principle

$\quad$ $C1$. $\quad q \cdot q \rightarrow. \, p \rightarrow p,$

the fallacy

$\quad$ $C2$. $\quad p \cdot p \rightarrow. \, p \rightarrow p \rightarrow p \cdot p,$

and the quite unmotivated

$\quad$ $C3$. $\quad (\sim p \lor. \, p \rightarrow p \rightarrow p) \lor. \, p \rightarrow p \lor p \rightarrow p \rightarrow. \, \mathrm{p}$

And though it lacks, on pain of inconsistency, both $p \rightarrow p^{2n}$ and $p^{2n} \rightarrow p$ for $n \geq 1$, where

- $p^1 = p$

- $p^{n+1} = (p^n \cdot p)$

it contains the peculiar principles

$\quad$ $C4$. $\quad p \rightarrow p^{2n+1}$ and
$\quad$ $C5$. $\quad p^{2n+1} \rightarrow p$, for $n \geq 0$

$\quad$ In all these respects, at any rate, the connexive logics that Angell has prepared *look* superior; and Angell indeed claims in [1] that none of $C1$–$C3$ are derivable in his systems PA1 and PA1m. Likewise he asserts that $p \rightarrow (p \cdot p) \cdot p$, *i.e.*, $C4$ for $n = 1$, is demonstrably not derivable in PA1 or PA1m. However, so far as we are aware, these non-derivability claims have never been made good. An advantage of the modellings we offer is that they enable the proof of some of these non-derivability results, and put us on the way to separating and rounding out axiomatically various different connexive logics.

Angell's system PA1m (which is supposed to be a modal version of the system PA1 of [2]) has the same syntax as CC1, with $A \supset B =_{Df} \sim(A \cdot \sim B)$, and these axioms:

A1.	$p \to p$	A2.	$q \cdot p \supset p$
A3.	$p \supset p \cdot p$	A4.	$p \cdot (q \cdot r) \to q \cdot (p \cdot r)$
A5.	$(r \cdot p) \cdot \sim(q \cdot r) \to p \cdot \sim q$	A6.	$\Box p \to p$
A7.	$p \to q \to . \Box p \to \Box q$	A8.	$p \to q \to \sim(p \to \sim q)$

(Minor changes have been made to A6 and A7, since $\Box$ is taken as primitive rather than, as in [1], $\Diamond$.) The rules are Detachment (for $\to$), Adjunction, and Substitution.

A CM1-*model* M is a structure $M = \langle G, K, I, R, v \rangle$ where $I \subseteq K$, $G \in U = K - I$, R is a binary relation on K and v is a bivalent valuation function on sentential parameters and elements of K, such that

i) R is transitive and reflexive;

ii) if $H_1 R H_2$ then $H_1 = H_2$ for $H_1 \in I$, *i.e.*, I-worlds are terminal

iii) if GRH then, for some $H_1 \in I$, HRH_1 (or equivalently, every world is R-connected to an I-world).

A CM1S-*model* is a CM1 model such that $I = \{J\}$, *i.e.* there is just one irregular world J.

The recursive extension of v to all wff in connective set $\{\sim, \cdot, \Box\}$ is the same as for CC1 except that the rule for $\cdot$ is generalized to:

- $v(A \cdot B, H) = T$ iff $v(A, H) = v(B, H)$ and either $H \in I$ or else $v(A, H) = T$.

A wff A is CM1-*valid* iff $v(A, G) = T$ for every CM1-model, *i.e.*, A is true in every CM1-model.

Theorem 3. *If B is a theorem of* PA1m *then B is* CM1-*valid.*

Proof. By induction over the length of proofs. Validation of some of the axioms, especially A7 and A8, requires the investigation of several cases according as $H \in I$ or $H \in U$. □

Corollary. *Neither $C2$ nor $C3$ is a theorem of* PA1m. *Hence* PA1m *is a proper subsystem of* CC1.

Proof. Proof is a matter of constructing countermodels to $C2$ and $C3$. We illustrate the method, familiar from modal logic semantics, by setting out a countermodel to $C2$. Let $K = \{G, H_1, H_2, J\}$; $I = \{J\}$; GRH_1, H_1RH_2, GRH_2 and HRJ and HRH for every $H \in K$; $v(p, H_1) = T$, $v(p, H_2) = F$, and otherwise v may be arbitrary, so say $v(S, H) = F$ otherwise. Since $v(p, J) \neq v(\sim p, J)$, $v(p \cdot \sim p, J) = F$ and $v(\sim(p \cdot \sim p), J) = T$. Since $v(p, H_2) = F$, $v(\sim p, H_2) = T$, whence $v(p \cdot \sim p, H_2) = F$ since $H_2 \in U$, and $v(\sim(p \cdot \sim p), H_2) = T$. Combining these evaluations $v(\Box \sim(p \cdot \sim p), H_2) = T$, *i.e.* $v(p \to p, H_2) = T$. Hence too $v(p \to p \cdot \sim(p \cdot p), H_1) = F$. But $v(p \cdot p, H_1) = T$, whence since GRH_1, $v(C2, G) = F$. $\qquad\square$

The converse of Theorem 3 is however false. Both $C1$ and the S4 axiom

$C6. \quad \Box p \to \Box\Box p$

are validated by CM1-models, though they are not theorems of PA1m. This can be shown by tightening up CM1-models in the way Kripke tightened up S4 models (in [3]) to provide a semantics for Lewis system S3.

A CM2-*model* M is a structure $\langle G, K, N, I, R, v \rangle$ which differs from a CM1 model only in having a set $N \subseteq K$ such that $G \in N$ and $I \subseteq N$. (It is enough also to require that R be a quasi-ordering on N.)

Theorem 4. *If B is a theorem of* PA1m *then B is* CM1-*valid.*

Corollary. *Neither $C1$ nor $C6$ is a theorem of* PA1m.

But again the converse of Theorem 4 is false. $C4$ and $C5$ are CM1-valid but neither appears to be a theorem of PA1m for any $n > 0$. This can be seen by modifying the rule for $\cdot$ in irregular worlds so that $\cdot$ behaves not like material equivalence but like some brand of *strict* equivalence. Since $p \leftrightarrow. (p \leftrightarrow p) \leftrightarrow p$, $p \leftrightarrow. (((p \leftrightarrow p) \leftrightarrow p) \leftrightarrow p)$, *etc.*, are not theorems of strict implication, I situations will serve then to falsify $C4$ and $C5$. The argument can be filled out by adding a relation S on K to the model, with S reflexive and transitive (and possibly coinciding with R), and by modifying the rule for $\cdot$ to the following:

- $v(A \cdot B, H) = T$ iff $v(A, H) = T = v(B, H)$, for $H \in U$; and

- $v(A \cdot B, H) = T$ iff for some $H_1 \in I$ if HSH_1 then $v(A, H_1) = v(B, H_1)$, for $H \in I$.

As well the modelling conditions have to be amended, condition ii) for CM1-models has to be abandoned, the requirements that for some $H_1 \in I$ HSH_1, for every $H \in I$ and that I is hereditary under S (*i.e.*, if $H_1 \in I$ and H_1SH_2 then $H_2 \in I$) added, *etc.*

References

[1] R. B. Angell. Connexive implication, modal logic and subjunctive conditionals. Unpublished manuscript.

[2] R. B. Angell. A propositional logic with subjunctive conditionals. *Journal of Symbolic Logic*, 27(3):327–343, 1962.

[3] S. A. Kripke. Semantical analysis of modal logic II. Non-normal modal propositional calculi. In *The Theory of Models*, pages 206–220. North-Holland Publishing Company, Amsterdam, 1965.

[4] E. J. Lemmon. Algebraic semantics for modal logics II. *Journal of Symbolic Logic*, 31(2):191–218, 1966.

[5] S. McCall. Connexive implication. *Journal of Symbolic Logic*, 31(3):415–433, 1966.

[6] R. Routley and R.K. Meyer. Every sentential logic has a two-valued worlds semantics. *Logique et Analyse*, 19(74–76):345–365, 1976.

[7] R. Routley and H. Montgomery. On systems containing Aristotle's Thesis. *Journal of Symbolic Logic*, 33(1):82–96, 1968.

Received posthumously

On a New Three-Valued Paraconsistent Logic

Grigory Olkhovikov
Ural Federal University, Russia
grigory.olkhovikov@gmail.com

Translated by
Thomas Macaulay Ferguson
Department of Philosophy
City University of New York Graduate Center, United States
tferguson@gradcenter.cuny.edu

1   Truth-Value Gaps and Interpreting Conditionals

Classically, the formulae of propositional logic are interpreted as follows: Variables accept values from the domain $\{0, 1\}$, in which 0 is the designated value and is interpreted as the truth of a corresponding expression, while 1 is interpreted as its falsity. The values of complex formulae are given by the following matrices:

p	q	$p \,\&\, q$	$p \vee q$	$\neg p$	$p \supset q$
0	0	0	0	1	0
0	1	1	0	1	1
1	0	1	0	0	0
1	1	1	1	0	0

In the sequel, we will refer to classical propositional logic, which is based on this interpretation of propositional formulae, as CL and the above matrices as C-matrices.

Translator's Note: The following is a translation of the paper *Ob odnoy novoy tryohznachnoy paraneprotivorechivoy logike*, which appeared in Russian as [3]. Much of the terminology of the original paper is drawn from a Russian translation of Alonzo Church's [1]; in such cases, the present translation employs Church's terminology whenever possible.

However, not all natural language propositions that are identified with true formulae of CL are true in their own right. In natural language, many propositions do not have definite truth values; in these cases we have *truth value gaps*. A prime example of such propositions is the case of conditional sentences with false antecedents. For the sake of simplicity of logical theory, such propositions are interpreted in CL as true and *prima facie* this correction of natural language appears justified. Arguing against the representation of such propositions as theorems of the theories we intend to formulate in CL seems equally impossible. However, these considerations do not negate the fact that the truth of those propositions—interpreted as true for the sake of formalization—differ radically from the truth of those propositions that are already true in natural language. The propositions in this latter category *should* be evaluated as true if we wish to obtain a correct formalization; the validity of their translations into formal languages is a necessary condition on the adequacy of a translation. However, there is no need to translate those propositions of natural language lacking a truth value as true; we may just as easily assign the value of "false" to all such propositions. That such formulae are treated as true, therefore, is arbitrary—or merely accidental—in contrast to the evaluation of the formal language counterparts of true natural language statements as essentially or necessarily true. The lack of a distinction between accidental and essential truth in the framework of CL disrupts the uniformity of the interpretation of truth-value gaps, as the sentences "If $2 + 2 = 5$ then $4 + 8 = 7$" and "It is not the case that if $2 + 2 = 5$ then $4 + 8 = 7$" do not have a particular truth value in natural language. But the first of them will be represented in CL as a true formula and the second, as false. In particular, such confusion leads to discussions concerning the so-called "paradoxes of material implication" (such as the law of affirmation of the consequent), although the proper reading of these formulae is hardly paradoxical.

In order to remedy the situation, I recommend constructing a propositional logic approximating the logical connectives of natural language as follows. We will employ the three-element set of truth values $\{0, 1, 2\}$ where 0 and 1 are designated values. 0 is understood as an essential truth, 1 as an accidental truth (or as truth for purposes of formalization) and 2 as the falsity of the expressions to which these values are assigned. It is supposed that the propositions of natural language that lack a definite truth value will be assigned a value of 1 when translated into the language of propositional logic. The truth values of complex formulae are specified by the following matrices (which will be called M-matrices):

p	q	$\neg p$	Lp	$p \vee q$	$p \,\&\, q$	$p \supset q$
0	0	2	0	0	0	0
0	1	2	0	0	1	1
0	2	2	0	0	2	2
1	0	1	2	0	1	0
1	1	1	2	1	1	1
1	2	1	2	1	2	2
2	0	0	2	0	2	1
2	1	0	2	1	2	1
2	2	0	2	2	2	1

Here, Lp means "p is essentially true" (or "p is necessarily true") and L can be regarded as a (truth-functional) strong alethic modality.

The formulae of propositional modal logic to which the M-matrices assign values from the set $\{0,1\}$ for all assignments of values to propositional variables will be called M-tautologies. There is an immediate connection between the interpretation of the connectives $\neg$, $\&$, and $\vee$ induced by the M-matrices and Kleene's three-valued logic K_3.[1] However, the definition of the implication connective that precisely captures the foregoing ideas on the truth value of the conditional in natural language is sufficiently novel and—to the best of my knowledge—has not been previously considered. In the sequel, the logic based on the interpretation of modal propositional formulae induced by M-matrices will be denoted by $LImp$.

One can raise the question of the functional completeness of the set of connectives $\{\neg, L, \&, \vee, \supset\}$. The answer to this question is yes, although we will not discuss the matter in detail. Rather, we will consider the grounds for stronger assertion of the functional completeness of the set $\{\neg, L, \supset\}$; an affirmative answer to the latter assertion will lead to an affirmative answer to the first issue. As is well-known,[2] any function of three-valued logic is expressible by compositions of the so-called Webb function $V_3(x,y) = \max(x,y) + 1$, where $+$ represents addition modulo 3. Hence, if we could define $V_3(x,y)$ as a composition of the functions $\neg$, L, $\supset$, this would constitute proof of the functional completeness of the set $\{\neg, L, \supset\}$. For example, one composition corresponding to the function $V_3(x,y)$ is given by the formula $\neg L(Lx \supset y) \supset \neg L \neg(x \supset L \neg y)$ ($\star$), as can be verified on the basis of the following table:

---

[1] On K_3, see A.S. Karpenko's *Logic and Computer* [2, 22-23].
[2] See S.V. Yablonsky's *Introduction to Discrete Mathematics* [5, 50].

x	y	$\max(x,y)$	$\max(x,y)+1$	Lx	$L\neg y$	$Lx \supset y$
0	0	0	1	0	2	0
0	1	1	2	0	2	1
0	2	2	0	0	0	2
1	0	1	2	2	2	1
1	1	1	2	2	2	1
1	2	2	0	2	0	1
2	0	2	0	2	2	1
2	1	2	0	2	2	1
2	2	2	0	2	0	1

x	y	$x \supset L\neg y$	$\neg L(Lx \supset y)$	$\neg L \neg(x \supset L\neg y)$	$(\star)$
0	0	2	2	2	1
0	1	2	0	2	2
0	2	0	0	0	0
1	0	2	0	2	2
1	1	2	0	2	2
1	2	0	0	0	0
2	0	1	0	0	0
2	1	1	0	0	0
2	2	1	0	0	0

Thus, the set $\{\neg, L, \supset\}$ is functionally complete, that is, any of the functions of three-valued logic can be expressed as compositions of instances of its members, including the other functions defined by the above M-matrices, *i.e.*, & and $\vee$. Hence, we take the connectives $\neg$, L, and $\supset$ to be the primitive connectives of *LImp*.

The set of M-tautologies can be determined in a purely syntactical fashion, *i.e.*, by means of an uninterpreted modal propositional calculus, in which all and only M-tautologies are deducible. But before we proceed to the examination of such a system, let us observe a number of important properties of *LImp*, which are already quite apparent at the semantic level. Firstly, all tautologies of *CL* that contain no instances of the symbol $\neg$ are M-tautologies. Secondly, the set of M-tautologies does not include all of the tautologies of *CL*. Furthermore, the set of M-tautologies includes neither the set of theorems of intuitionistic propositional logic nor the set of theorems of the Kolmogorov-Johansson minimal propositional logic. Indeed, the formula $(p \supset q) \supset ((p \supset \neg q) \supset \neg p)$ is provable in *CL* as well as the intuitionistic and minimal propositional logics but is not an M-tautology. Thirdly, the set of M-tautologies in which no instances of the modal operator L appear is not exhausted by the set of formulae provable in *CL*. Indeed, the M-tautology $(p \supset \neg q) \supset \neg(p \supset q)$ cannot be proved in *CL*. It is also clear that *LImp* is a weakly paraconsistent logic, because the formula $p \,\&\, \neg p$—although not an M-tautology—turns out to be satisfiable according to the M-matrices. Moreover, it is easy to see that the negation of the "law of excluded third," *i.e.*, the formula $\neg(p \vee \neg p)$, is also satisfiable. Finally, if we consider *LImp* as a modal logic, we find that in this respect its characteristics are quite interesting. It is a non-normal modal logic insofar as it does not admit Gödel's rule. On the other hand, as we shall see below, many intuitively correct theorems about modality are provable in *LImp* (see theorems T13, T17, T18 below). *LImp* does not verify the well-known formulae that express the so-called "paradoxes of strict implication," that is, the formulae $L\,p \supset L(q \supset p)$ and $L\,\neg p \supset L(p \supset q)$, although the no less paradoxical "converse" formulae $L\,\neg p \supset \neg(p \supset q)$ and $L(q \supset p) \supset L\,p$ are M-tautologies. In contrast to the four-valued logic of Łukasiewicz, in *LImp* there exist theorems of the form $L\,\alpha$, such as, for example, the formula $L\,\neg L\,\neg(p \supset p)$. Hence, *LImp*, construed as a modal logic, is intuitive in a number of respects, especially in virtue of its relative simplicity. Of course, it is not a perfect modal logic and it is very easy to point out implausible and even paradoxical M-tautologies.

To my mind, the following definition of modalities will be most productive within the framework of *LImp*, which differs from the traditional definition in several ways, in particular by excluding the so-called improper modalities of p and $\neg p$. Modalities in *LImp* will be assumed to be any unary truth function that maps the set $\{0, 1, 2\}$ into the set $\{0, 2\}$. From this perspective, there are only eight pairwise distinct modalities, defined in particular by the following formulae of *LImp*: $L(p \supset \neg p)$, $\neg L(p \supset \neg p)$, $L\,p$, $L\,\neg p$, $\neg L\,p$, $\neg L\,\neg p$, $L\,\neg(p \supset L\,p)$, $\neg L\,\neg(p \supset L\,p)$.

2 *LImp* as an Axiomatic System

The primitive symbols of *LImp* are the propositional variables that appear in the following countably infinite enumeration:

$$p, q, r, s, p_1, q_1, ...,$$

In addition, the primitive symbols include the unary connectives L and $\neg$ and binary connective $\supset$. The definitions of a well-formed formula, proofs, and deriving a conclusion from hypotheses are given in the standard way. In the metalanguage, the statement $\vdash \alpha$ means that α is provable in *LImp* while $\alpha_1, ..., \alpha_n \vdash \beta$ means that β is provable in *LImp* from hypotheses $\alpha_1, ..., \alpha_n$.

The *axioms* of *LImp* are:

A1. $p \supset (q \supset p)$

A2. $(p \supset (q \supset r)) \supset ((p \supset q) \supset (p \supset r))$

A3. $\neg(p \supset q) \supset (p \supset \neg q)$

A4. $(\neg p \supset \neg\neg p) \supset p$

A5. $(Lp \supset L\neg q) \supset \neg(q \supset p)$

A6. $(\neg p \supset \neg q) \supset (Lq \supset L\neg\neg Lp)$

The *inference rules* of $LImp$ are:

R1. Rule of substitution: Substitute formulae $\alpha_1, ..., \alpha_n$ for propositional variables $\beta_1, ..., \beta_n$, respectively.

R2. Modus ponens (MP): If $\vdash A$ and $\vdash A \supset B$, then $\vdash B$

In the sequel, a substitution instance of a previously proven formula F is denoted $F(\alpha_1/\beta_1, ..., \alpha_n/\beta_n)$. If a provable formula appears on the mth line of the proof of a theorem T^*, we will consider it to be a theorem as well and will denote it as $T^*: m$. For an inference rule R differing from the rule of substitution, either primitive or derived, the application of R to formulae appearing at lines $m_1, ..., m_n$ in a derivation will be denoted by $R(m_1, ..., m_n)$.

Just as in CL, by appeal to $A1$ and $A2$ the following *theorems* can be proven:

T1. $p \supset p$

T2. $(p \supset (p \supset q)) \supset (p \supset q)$

T3. $p \supset ((p \supset q) \supset q)$

T4. $(p \supset (q \supset r) \supset (q \supset (p \supset r))$

T5. $(p \supset q) \supset ((q \supset r) \supset (p \supset r))$

T6. $(q \supset r) \supset ((p \supset q) \supset (p \supset r))$

T7. $((p \supset q) \supset (p \supset r)) \supset (p \supset (q \supset r))$

And the following derived rules of inference can be shown to be admissible:

DR1. $A \supset B, B \supset C \vdash A \supset C$

DR2. $A \supset B \vdash (B \supset C) \supset (A \supset C)$

DR3. $B \supset C \vdash (A \supset B) \supset (A \supset C)$

DR4. $A \supset (B \supset C) \vdash B \supset (A \supset C)$

DR5. If $\vdash A \supset B$ and $\vdash B \supset C$ then $\vdash A \supset C$

DR6. If $\vdash A \supset B$ then $\vdash (B \supset C) \supset (A \supset C)$

DR7. If $\vdash B \supset C$ then $\vdash (A \supset B) \supset (A \supset C)$

DR8. If $\vdash A \supset (B \supset C)$ then $\vdash B \supset (A \supset C)$

DR9. $A_1, ..., A_{i-1}, A_i, ..., A_n \vdash B$ if and only if
$$A_1, ..., A_{i-1} \vdash A_i \supset (... \supset (A_n \supset B)) \text{ for } 1 \leq i \leq n$$

Now, we prove some important theorems of $LImp$ that will be needed in the sequel.

T8. $(Lp \supset L\neg q) \supset (q \supset \neg p)$

1	$(Lp \supset L\neg q) \supset \neg(q \supset p)$	$A5$
2	$\neg(q \supset p) \supset (q \supset \neg p)$	$A3(q/p, p/q)$
3	$(Lp \supset L\neg q) \supset (q \supset \neg p)$	$DR5(1,2)$

T9. $p \supset \neg\neg p$

1	$L\neg p \supset L\neg p$	$T1(L\neg p/p)$
2	$(L\neg p \supset L\neg p) \supset (p \supset \neg\neg p)$	$T8(\neg p/p, p/q)$
3	$p \supset \neg\neg p$	$MP(1,2)$

T10. $\neg\neg p \supset p$

1	$\neg\neg p \supset (\neg p \supset \neg\neg p)$	$A1(\neg\neg p/p, \neg p/q)$
2	$(\neg p \supset \neg\neg p) \supset p$	$A4$
3	$\neg\neg p \supset p$	$DR5(1,2)$

T11. $L\neg p \supset (p \supset q)$

1	$L\neg p \supset (L\neg q \supset L\neg p)$	$A1(L\neg p/p, L\neg q/q)$
2	$(L\neg q \supset L\neg p) \supset (p \supset \neg\neg q)$	$T8(\neg q/p, p/q)$
3	$\neg\neg q \supset q$	$T10(q/p)$
4	$(p \supset \neg\neg q) \supset (p \supset q)$	$DR7(3)$
5	$(L\neg q \supset L\neg p) \supset (p \supset q)$	$DR5(2,4)$
6	$L\neg p \supset (p \supset q)$	$DR5(1,5)$

T12. $L\neg L\neg p \supset p$

1	$L\neg L\neg p \supset (L\neg p \supset L\neg(p \supset p))$	$T11(L\neg p/p, L\neg(p \supset p)/q)$
2	$(L\neg p \supset L\neg(p \supset p)) \supset ((p \supset p) \supset p)$	$T11:5(p \supset p/p, p/q)$
3	$L\neg L\neg p \supset ((p \supset p) \supset p)$	$DR5(1,2)$
4	$(p \supset p) \supset (L\neg L\neg p \supset p)$	$DR8(3)$
5	$p \supset p$	$T1$
6	$L\neg L\neg p \supset p$	$MP(4,5)$

T13. $p \supset L\neg L\neg p$

1	$L\neg L\neg L\neg p \supset L\neg p$	$T12(L\neg p/p)$
2	$(L\neg L\neg L\neg p \supset L\neg p) \supset (p \supset L\neg L\neg p)$	$T11:5(L\neg L\neg p/q)$
3	$p \supset L\neg L\neg p$	$MP(1,2)$

T14. $(\neg p \supset \neg q) \supset (Lq \supset Lp)$

1	$L\neg\neg Lp \supset (\neg Lp \supset \neg\neg Lp)$	$T11(\neg Lp/p, \neg\neg Lp/q)$
2	$(\neg Lp \supset \neg\neg Lp) \supset Lp$	$A4(Lp/p)$
3	$L\neg\neg Lp \supset Lp$	$DR5(1,2)$
4	$(\neg p \supset \neg q) \supset (Lq \supset L\neg\neg Lp)$	$A6$
5	$(Lq \supset L\neg\neg Lp) \supset (Lq \supset Lp)$	$DR7(3)$
6	$(\neg p \supset \neg q) \supset (Lq \supset Lp)$	$DR5(4,5)$

T15. $L\neg\neg p \supset Lp$

1	$\neg p \supset \neg\neg\neg p$	$T9(\neg p/p)$
2	$(\neg p \supset \neg\neg\neg p) \supset (L\neg\neg p \supset Lp)$	$T14(\neg\neg p/q)$
3	$L\neg\neg p \supset Lp$	$MP(1,2)$

T16. $Lp \supset L\neg\neg p$

1	$\neg\neg\neg p \supset \neg p$	$T10(\neg p/p)$
2	$(\neg\neg\neg p \supset \neg p) \supset (Lp \supset L\neg\neg p)$	$T14(\neg\neg p/p, p/q)$
3	$Lp \supset L\neg\neg p$	$MP(1,2)$

T17. $Lp \supset LLp$

1	$\neg p \supset \neg p$	$T1(\neg p/p)$
2	$(\neg p \supset \neg p) \supset (Lp \supset L\neg\neg Lp)$	$A6(p/q)$
3	$Lp \supset L\neg\neg Lp$	$MP(1,2)$
4	$L\neg\neg Lp \supset LLp$	$T15(Lp/p)$
5	$Lp \supset LLp$	$DR5(3,4)$

T18. $Lp \supset p$

1	$L\neg\neg p \supset (\neg p \supset \neg\neg p)$	$T11(\neg p/p, \neg\neg p/q)$
2	$Lp \supset L\neg\neg p$	$T16$
3	$Lp \supset (\neg p \supset \neg\neg p)$	$DR5(1,2)$
4	$(\neg p \supset \neg\neg p) \supset p$	$A4$
5	$Lp \supset p$	$DR5(3,4)$

T19. $(p \supset q) \supset (L\,\neg q \supset L\,\neg p)$

1	$p \supset q$	Premise
2	$q \supset \neg\neg q$	$T9(q/p)$
3	$(p \supset q) \supset (p \supset \neg\neg q)$	$DR3(2)$
4	$p \supset \neg\neg q$	$MP(1,3)$
5	$\neg\neg p \supset p$	$T10$
6	$(p \supset \neg\neg q) \supset (\neg\neg p \supset \neg\neg q)$	$DR2(5)$
7	$\neg\neg p \supset \neg\neg q$	$MP(4,6)$
8	$(\neg\neg p \supset \neg\neg q) \supset (L\,\neg q \supset L\,\neg p)$	$T14(\neg p/p, \neg q/q)$
9	$L\,\neg q \supset L\,\neg p$	$MP(7,8)$
10	$(p \supset q) \supset (L\,\neg q \supset L\,\neg p)$	$DR9(1\text{--}9)$

T20. $\neg p \supset L\,\neg L\,p$

1	$\neg p \supset L\,\neg L\,\neg\neg p$	$T13(\neg p/p)$
2	$L\,p \supset L\,\neg\neg p$	$T16$
3	$(L\,p \supset L\,\neg\neg p) \supset (L\,\neg L\,\neg\neg p \supset L\,\neg L\,p)$	$T19(L\,p/p, L\,\neg\neg p/q)$
4	$L\,\neg L\,\neg\neg p \supset L\,\neg L\,p$	$MP(2,3)$
5	$\neg p \supset L\,\neg L\,p$	$DR5(1,4)$

T21. $(p \supset \neg q) \supset \neg(p \supset q)$

1	$(p \supset \neg q) \supset (L\,\neg\neg q \supset L\,\neg p)$	$T19(\neg q/q)$
2	$L\,q \supset L\,\neg\neg q$	$T16(q/p)$
3	$(L\,\neg\neg q \supset L\,\neg p) \supset (L\,q \supset L\,\neg p)$	$DR6(2)$
4	$(L\,q \supset L\,\neg p) \supset \neg(p \supset q)$	$A5(q/p, p/q)$
5	$(L\,\neg\neg q \supset L\,\neg p) \supset \neg(p \supset q)$	$DR5(3,4)$
6	$(p \supset \neg q) \supset \neg(p \supset q)$	$DR5(1,5)$

T22. $(L\,\neg p \supset p) \supset p$

1	$L\,\neg p \supset (p \supset L\,\neg(p \supset p))$	$T11(L\,\neg(p \supset p)/q)$
2	$(L\,\neg p \supset (p \supset L\,\neg(p \supset p))) \supset$ $((L\,\neg p \supset p) \supset (L\,\neg p \supset L\,\neg(p \supset p)))$	$A2(L\,\neg p/p, p/q,$ $L\,\neg(p \supset p)/r)$
3	$(L\,\neg p \supset p) \supset (L\,\neg p \supset L\,\neg(p \supset p))$	$MP(1,2)$
4	$(L\,\neg p \supset L\,\neg(p \supset p)) \supset ((p \supset p) \supset p)$	$T11{:}\,5(p \supset p/p, p/q)$
5	$(L\,\neg p \supset p) \supset ((p \supset p) \supset p)$	$DR5(3,4)$
6	$(p \supset p) \supset ((L\,\neg p \supset p) \supset p)$	$DR8(5)$
7	$p \supset p$	$T1$
8	$(L\,\neg p \supset p) \supset p$	$MP(6,7)$

T23. $(Lp \supset q) \supset ((p \supset (\neg p \supset q)) \supset ((L \neg p \supset q) \supset q))$

1	$Lp \supset q$	Premise
2	$p \supset (\neg p \supset q)$	Premise
3	$L \neg p \supset q$	Premise
4	$(L \neg p \supset q) \supset (L \neg q \supset L \neg L \neg p)$	$T19(L \neg p/p)$
5	$L \neg q \supset L \neg L \neg p$	$MP(3,4)$
6	$L \neg L \neg p \supset p$	$T12$
7	$L \neg q \supset p$	$DR1(5,6)$
8	$L \neg q \supset (\neg p \supset q)$	$DR1(2,7)$
9	$(\neg p \supset q) \supset (L \neg q \supset L \neg\neg p)$	$T19(\neg p/p)$
10	$L \neg q \supset (L \neg q \supset L \neg\neg p)$	$DR1(8,9)$
11	$(L \neg q \supset (L \neg q \supset L \neg\neg p)) \supset$ $(L \neg q \supset L \neg\neg p)$	$T2(L \neg q/p, L \neg\neg p/q)$
12	$L \neg q \supset L \neg\neg p$	$MP(10,11)$
13	$L \neg\neg p \supset Lp$	$T15$
14	$L \neg q \supset Lp$	$DR1(12,13)$
15	$L \neg q \supset q$	$DR1(1,14)$
16	$(L \neg q \supset q) \supset q$	$T22(q/p)$
17	q	$MP(15,16)$
18	$(Lp \supset q) \supset ((p \supset (\neg p \supset q)) \supset$ $((L \neg p \supset q) \supset q))$	$DR9(1\text{--}17)$

T24. $L \neg p \supset \neg(p \supset q)$

1	$L \neg p \supset (p \supset \neg q)$	$T11(\neg q/q)$
2	$(p \supset \neg q) \supset \neg(p \supset q)$	$T21$
3	$L \neg p \supset \neg(p \supset q)$	$DR5(1,2)$

T25. $p \supset (L \neg q \supset L \neg(p \supset q))$

1	$p \supset ((p \supset q) \supset q)$	$T3$
2	$((p \supset q) \supset q) \supset (L \neg q \supset L \neg(p \supset q))$	$T19(p \supset q/p)$
3	$p \supset (L \neg q \supset L \neg(p \supset q))$	$DR5(1,2)$

T26. $p \supset (Lq \supset L(p \supset q))$

1	$\neg(p \supset q) \supset (p \supset \neg q)$	$A3$
2	$p \supset (\neg(p \supset q) \supset \neg q)$	$DR8(1)$
3	$(\neg(p \supset q) \supset \neg q) \supset (Lq \supset L(p \supset q))$	$T14(p \supset q/p)$
4	$p \supset (Lq \supset L(p \supset q))$	$DR5(2,3)$

T27. $\neg q \supset \neg(p \supset q)$

1	$\neg q \supset (p \supset \neg q)$	$A1(\neg q/q)$
2	$(p \supset \neg q) \supset \neg(p \supset q)$	$T21$
3	$\neg q \supset \neg(p \supset q)$	$DR5(1,2)$

The foregoing theorems allow us to prove the following *metatheorems* that highlight important properties of the system *LImp*.

Metatheorem 1. *Let A be a formula of the system $LImp$ and let $p_0, ..., p_n$ be a sequence of propositional variables including each variable appearing in A. Let $t_1, ..., t_n$ be some values of $\{0, 1, 2\}$ corresponding to $p_1, ..., p_n$ and let $\alpha(t_1, ..., t_n)$ be a sequence of formulae satisfying the following conditions:*

1. *If $t_i = 0$ then $L\, p_i \in \alpha(t_1, ..., t_n)$*

2. *If $t_i = 1$ then $p_i \in \alpha(t_1, ..., t_n)$ and $\neg p_i \in \alpha(t_1, ..., t_n)$*

3. *If $t_i = 2$ then $L \neg p_i \in \alpha(t_1, ..., t_n)$*

4. *A formula is included in the sequence $\alpha(t_1, ..., t_n)$ only in virtue of clauses 1–3*

5. *If $i < j$, then any formula containing p_i appears at an earlier stage in the sequence $\alpha(t_1, ..., t_n)$ than any formula containing p_j*

6. *p_i appears in the sequence $\alpha(t_1, ..., t_n)$ at an earlier stage than $\neg p_i$*

Then the following assertions hold with respect to the M-matrices:

I If the values $t_1, ..., t_n$ are assigned to variables $p_1, ..., p_n$, respectively, the formula A receives a value of 0, then $\alpha(t_1, ..., t_n) \vdash L\, A$.

II If the values $t_1, ..., t_n$ are assigned to variables $p_1, ..., p_n$, respectively, the formula A receives a value of 1, then $\alpha(t_1, ..., t_n) \vdash A$ and $\alpha(t_1, ..., t_n) \vdash \neg A$.

III If the values $t_1, ..., t_n$ are assigned to variables $p_1, ..., p_n$, respectively, the formula A receives a value of 2, then $\alpha(t_1, ..., t_n) \vdash L \neg A$.

Proof. We prove Metatheorem 1 by induction on $s(A)$, the number of occurrences of the symbols L, $\neg$, and $\supset$ in the formula A.

Basis step: Let $s(A) = 0$. Then trivially we have that $L\, p_i \vdash L\, p_i$ holds, that both $p_i, \neg p_i \vdash p_i$ and $p_i, \neg p_i \vdash \neg p_i$ hold, and that $L \neg p_i \vdash L \neg p_i$ holds, whence assertions I–III are demonstrated to hold.

Induction step: Suppose that whenever $s(A) \leq k$, assertions I–III hold and let $s(A) = k + 1$. Then there are precisely three cases:

1. A has the form $A' \supset A''$, where $s(A') \leq k$ and $s(A'') \leq k$.

1.1. If the value assigned to A for values $t_1, ..., t_n$ according to the M-matrices is 0, then for the same values $t_1, ..., t_n$, A'' receives a value of 0 while A' receives a value from $\{0, 1\}$. By induction hypothesis, we have:

$$\alpha(t_1, ..., t_n) \vdash L\,A'' \tag{1}$$

and either:

$$\alpha(t_1, ..., t_n) \vdash A' \tag{2}$$

or:

$$\alpha(t_1, ..., t_n) \vdash L\,A' \tag{3}$$

If (3) holds, in virtue of $T18$ we infer that $\vdash L\,A' \supset A'$ and by applying MP again infer that $\alpha(t_1, ..., t_n) \vdash A'$. Hence, (2) holds in either case. By $T26$ we know that $\vdash A' \supset (L\,A'' \supset L(A' \supset A''))$, and by appeal to MP, from (2) we obtain the proposition that $\alpha(t_1, ..., t_n) \vdash L\,A'' \supset L(A' \supset A'')$. Thus, by an application of MP to (1), we infer that $\alpha(t_1, ..., t_n) \vdash L(A' \supset A'')$.

1.2. If the value assigned to A by the M-matrices under values $t_1, ..., t_n$ is 2, then for the same values $t_1, ..., t_n$ A'' takes the value 2 and A' takes a value from $\{0, 1\}$. Thus, by induction hypothesis, we infer that:

$$\alpha(t_1, ..., t_n) \vdash L\,\neg A'' \tag{4}$$

and, as before, are able to infer that either (2) holds or (3) holds. The above reasoning again establishes that we may derive (2) in either case. By $T25$, we have $\vdash A' \supset (L\,\neg A'' \supset L\neg(A' \supset A''))$. Hence, it follows from the application of MP to (2) that we obtain $\alpha(t_1, ..., t_n) \vdash L\,\neg A'' \supset L\,\neg(A' \supset A'')$. From this and (4), by MP we infer that $\alpha(t_1, ..., t_n) \vdash L\,\neg(A' \supset A'')$.

1.3. If the value of A under the assignment $t_1, ..., t_n$ according to the M-matrices is 1, then under the same values $t_1, ..., t_n$, either A'' receives a value of 1 or A' receives a value of 2. In the first case we have:

$$\alpha(t_1, ..., t_n) \vdash A'' \tag{5}$$

and

$$\alpha(t_1, ..., t_n) \vdash \neg A'' \tag{6}$$

As an instance of $A1$, we obtain $\alpha(t_1, ..., t_n) \vdash A'' \supset (A' \supset A'')$. From this in conjunction with (5), we derive $\alpha(t_1, ..., t_n) \vdash A' \supset A''$.

By $T27$ we obtain $\vdash \neg A'' \supset \neg(A' \supset A'')$. From this and (6) we infer that $\alpha(t_1, ..., t_n) \vdash \neg(A' \supset A'')$. In the latter case, we have:

$$\alpha(t_1, ..., t_n) \vdash L \neg A' \tag{7}$$

As an instance of $T11$, we have $\vdash L \neg A' \supset (A' \supset A'')$. From this in conjunction with (7) we may derive that $\alpha(t_1, ..., t_n) \vdash A' \supset A''$.

As an instance of $T24$, we obtain $L \neg A' \supset \neg(A' \supset A'')$. Together with (7), from this we infer that $\alpha(t_1, ..., t_n) \vdash \neg(A' \supset A'')$.

2. A has the form $\neg A'$, where $s(A') \leq k$.

2.1. If the M-matrices assign A a value of 0 under the values $t_1, ..., t_n$, then under the same values $t_1, ..., t_n$, A' takes the value of 2. Then, by induction hypothesis, we have:

$$\alpha(t_1, ..., t_n) \vdash L \neg A' \tag{8}$$

But $L \neg A'$ syntactically coincides with $L A$, so that (8) can also be written as the assertion that $\alpha(t_1, ..., t_n) \vdash L A$.

2.2. If the M-matrices ensure that the assignment of values $t_1, ..., t_n$ assigns A a value of 1, then for the same values $t_1, ..., t_n$, A' receives the value of 1 as well. By the induction hypothesis, we obtain:

$$\alpha(t_1, ..., t_n) \vdash \neg A' \tag{9}$$
$$\alpha(t_1, ..., t_n) \vdash A' \tag{10}$$

As an instance of $T9$, we have $A' \supset \neg\neg A'$. By applying MP to this and (10), we obtain $\alpha(t_1, ..., t_n) \vdash \neg\neg A'$. This last derivation can be rewritten in the form $\alpha(t_1, ..., t_n) \vdash \neg A$, while (9) can itself be rewritten as $\alpha(t_1, ..., t_n) \vdash A$.

2.3. When the M-matrices assign the value 2 to A under the assignment of values $t_1, ..., t_n$, then the values $t_1, ..., t_n$ also assign A' the value 0. By induction hypothesis, we have:

$$\alpha(t_1, ..., t_n) \vdash L A' \tag{11}$$

By appeal to $T16$, we infer that $\vdash L A' \supset L \neg\neg A'$, i.e., $\vdash L A' \supset L \neg A$. By an application of MP to this and (11), we obtain that $\alpha(t_1, ..., t_n) \vdash L \neg A$.

3. A is of the form $L\,A'$, where $s(A') \le k$.

3.1. Now, if under the values $t_1, ..., t_n$, A receives a value of 0 according to the M-matrices, then the same values $t_1, ..., t_n$ assign A' the value 0. By induction hypothesis, we infer that:

$$\alpha(t_1, ..., t_n) \vdash L\,A' \tag{12}$$

As an instance of $T17$, we obtain $L\,A' \supset L\,L\,A'$. By applying MP, from this and (12) we obtain $\alpha(t_1, ..., t_n) \vdash L\,L\,A'$.

3.2. If according to the M-matrices, under the assignment of values $t_1, ..., t_n$, A receives a value of 2, then these values $t_1, ..., t_n$ assign the formula A' a value from the set $\{1, 2\}$. By induction hypothesis, we know that either:

$$\alpha(t_1, ..., t_n) \vdash \neg A' \tag{13}$$

or:

$$\alpha(t_1, ..., t_n) \vdash L\,\neg A' \tag{14}$$

In the case in which (14) holds, then we have $\vdash L\,\neg A' \supset \neg A'$ as an instance of $T18$ and by applying MP to this and (14), we have (13) as well. Thus, (13) holds in either case. By $T20$, we have $\vdash \neg A' \supset L\,\neg L\,A'$. From this and (13), by applying MP we obtain $\alpha(t_1, ..., t_n) \vdash L\,\neg L\,A'$.

This proves Metatheorem 1.

$\square$

Metatheorem 2. *For all formulae A, if $\vdash A$, then A is an M-tautology.*

Proof. It is easy to confirm that each of the axioms of $LImp$ is an M-tautology and that the rules of inference of $LImp$ preserve the property of being an M-tautology.

$\square$

Metatheorem 3. *If a formula A is an M-tautology, then $\vdash A$.*

Proof. If A is an M-tautology, then for all $t_1, ..., t_n$, either

$$\alpha(t_1, ..., t_n) \vdash L\,A \tag{15}$$

or

$$\alpha(t_1, ..., t_n) \vdash A \tag{16}$$

where $\alpha(t_1, ..., t_n)$ is as defined as in Metatheorem 1. If (15) holds, then by $T18$, we have $\vdash L A \supset A$ and by MP obtain (16). Since we have (16) in both cases, by the definition of $\alpha(t_1, ..., t_n)$, for any $t_1, ..., t_{n-1}$ we infer correctness of the following inferences:

$$\alpha(t_1, ..., t_{n-1}), L p_n \vdash A \qquad (17)$$
$$\alpha(t_1, ..., t_{n-1}), p_n, \neg p_n \vdash A \qquad (18)$$
$$\alpha(t_1, ..., t_{n-1}), L \neg p_n \vdash A \qquad (19)$$

By $DR9$, from (17)–(19), we obtain:

$$\alpha(t_1, ..., t_{n-1}) \vdash L p_n \supset A \qquad (20)$$
$$\alpha(t_1, ..., t_{n-1}) \vdash p_n \supset (\neg p_n \supset A) \qquad (21)$$
$$\alpha(t_1, ..., t_{n-1}) \vdash L \neg p_n \supset A \qquad (22)$$

By $T23$ we have $\vdash (L p_n \supset A) \supset ((p_n \supset (\neg p_n \supset A)) \supset ((L \neg p_n \supset A) \supset A))$. From this in conjunction with (20)–(22), by three applications of MP we obtain $\alpha(t_1, ..., t_{n-1}) \vdash A$ for arbitrary $t_1, ..., t_{n-1}$. In this way, by successively reducing the index i in $\alpha(t_1, ..., t_i) \vdash A$ where $1 \leq i \leq n$, we can clearly ensure that $\vdash A$ holds. This proves Metatheorem 3. □

Before proceeding to the proofs of further metatheorems, we introduce a number of important concepts. A formal system is called absolutely consistent whenever there exists a formula that is not provable in the system. A formal system is called consistent in the sense of Post if and only if no formulae in which no connectives appear are provable in the system. A system is called relatively consistent with respect to a transformation mapping formulae A to formulae A^* if and only if there exists no formula A such that both A and A^* are theorems of that system. A formal system is called absolutely complete (alternately, complete in the sense of Post or with respect to a transformation mapping formulae A to A^*) if and only if the enrichment of that system by an arbitrary new axiom makes the system absolutely inconsistent (inconsistent in the sense of Post or with respect to a transformation mapping formulae A to A^*, respectively).

Metatheorem 4. *The system $LImp$ is:*

I Absolutely consistent.

II Consistent in the sense of Post.

III Relatively consistent with respect to the transformation mapping A to $L \neg A$.

IV Relatively inconsistent with respect to the transformation mapping A to $\neg A$.

Proof. For I, the property follows from Metatheorem 2 that any formula that is not M-tautologous is not provable in $LImp$.

For II, note that no formula in which no connectives appear (*i.e.*, an atomic formula) is an M-tautology. Hence, in virtue of Metatheorem 2, no such formula is provable in $LImp$.

For III, it follows from the M-matrices that for any formula A, it is not the case that both A and $L \neg A$ are M-tautologies. Hence, by Metatheorems 2 and 3, at least one of these formulae is unprovable in $LImp$.

For IV, consider the formulae $L \neg (p \supset p) \supset p$ and $\neg (L \neg (p \supset p) \supset p)$, both of which are M-tautologies. By appeal to Metatheorem 3, each of these formulae are provable in $LImp$. □

Metatheorem 5. *The system $LImp$ is:*

I Absolutely complete.

II Complete in the sense of Post.

III Complete relative to the transformation mapping formulae A to $L \neg A$.

Proof. Suppose that A is a formula unprovable in $LImp$. Then A is not an M-tautology, *i.e.*, there are values $t_1, ..., t_n$ that, when assigned to the atomic formulae $p_1, ..., p_n$ appearing in A, assign A the value of 2. Let A^* be the formula obtained from A by the following substitution:

- If $t_i = 0$, then in place of p_i, substitute the formula $\neg L \neg (p \supset p)$

- If $t_i = 1$, then in place of p_i, substitute the formula $L \neg (p \supset p) \supset p$

- If $t_i = 2$, then in place of p_i, substitute the formula $L \neg (p \supset p)$

If we add A to $LImp$ as a new axiom, then by an application of the rule of substitution, we infer that $\vdash A^*$. However, according to the M-matrices, the value assigned to A^* is identical to 2. Hence, the value assigned to $L \neg A^*$ is equal to 0, and—because $L \neg A^*$ is thus an M-tautology—the formula is provable in $LImp$. We have $\vdash A^*$, $\vdash L \neg A^*$, and—by appeal to $T11$—also $\vdash L \neg A^* \supset (A^* \supset B)$, where B is an arbitrary formula. By two applications of MP, we infer that $\vdash B$, *i.e.*, $LImp + A$ is absolutely inconsistent and, as a result, both inconsistent in the sense of Post and inconsistent with respect to the transformation mapping A to $L \neg A$. This proves Metatheorem 5. □

Metatheorem 6. *All axioms of $LImp$—except, perhaps, A1—are independent.*

Proof. 1) For the independence of $A2$, fix a set of truth values $\{0,1,2,3,4\}$ with 0 and 1 designated. Consider the following interpretation of the connectives of $LImp$:

p	$\neg p$	Lp
0	4	0
1	1	4
2	0	4
3	0	4
4	0	4

$\supset$	0	1	2	3	4
0	0	1	4	4	4
1	0	1	4	4	4
2	1	1	1	4	1
3	1	1	1	1	1
4	1	1	1	1	1

2) Let $\{0,1,2\}$ be a fixed set of truth values with 0 and 1 designated values and let functors L and $\neg$ be interpreted in the same manner as in the M-matrices. Then independence of $A3$ can be proven by considering the following interpretation of the connective $\supset$:

$\supset$	0	1	2
0	1	1	2
1	1	1	2
2	1	1	1

3) Let $\{0,1\}$ be a fixed set of truth values with 0 the designated value. Let $\supset$ be interpreted as in the C-matrices with L and $\neg$ interpreted according to the following table:

p	Lp	$\neg p$
0	0	0
1	1	0

The foregoing interpretation verifies all the axioms and rules of inference of $LImp$ except $A4$, thereby proving the independence of the latter.

4) Independence of $A5$ as an axiom of $LImp$ can be proven by appeal to the following interpretation of its formulae. Fix a set of truth values $\{0,1,2\}$ with 0 and 1 as its designated values. Let the functors $\supset$ and $\neg$ be interpreted in the same manner as the M-matrices and interpret L as the functor mapping all arguments to a value of 0.

5) Independence of $A6$ as an axiom of $LImp$ can be proven by using the following interpretation of its formulae. Let $\{0,1,2\}$ be a fixed set of truth values with 0 and 1 designated values. Let the functors $\supset$ and $\neg$ be interpreted as in the case of M-matrices and the functor L be interpreted according to the following array:

p	Lp
0	1
1	2
2	2

□

This proves Metatheorem 6.

References

[1] A. Church. *Introduction to Mathematical Logic*. Princeton University Press, Princeton, 1956.

[2] A.S. Karpenko. *Logic and Computer*, volume 4 of *Many-Valued Logics*. Nauka, Moscow, 1997. In Russian.

[3] G. K. Olkhovikov. On a new three-valued paraconsistent logic. In *Logic of Law and Tolerance*, pages 96–113, Yekaterinburg, 2002. Ural State University Press. In Russian.

[4] S.V. Yablonsky. *Introduction to Discrete Mathematics*. Mir Publishers, Moscow, 1989. O. Efimov, translator.

[5] S.V. Yablonsky. *Introduction to Discrete Mathematics*. Vysshaya Shkola, Moscow, 2001. In Russian; translated as [4].

Received in translation

A Complete, Correct, and Independent Axiomatization of the First-Order Fragment of a Three-Valued Paraconsistent Logic

Grigory Olkhovikov
Ural Federal University, Russia
grigory.olkhovikov@gmail.com

Abstract

I designed the three-valued paraconsistent logic $LImp$ as a very slight and natural improvement of classical material implication. From a commonsense point of view, both statements $(2+2) = 5 \to (0 = 0)$ and $\neg((2+2) = 5 \to (0 = 0))$ present a truth-value gap. However, within the classical two-valued logic, the first one is true, and the last one is false. Hence, I added the third (designated) value 'truth-value gap' and stipulated that an implication has this value if its antecedent is false. Otherwise, the implication truth value is the truth value of the implication consequent. If one adds the negation which maps truth-value gap to truth value gap and behaves classically in other cases, and the 'material truth' functor L mapping truth to truth and other values to falsehood, then a functionally complete set of three-valued connectives is obtained. I investigated the propositional logic operating with the described set of connectives and designed a complete, correct and independent axiomatization of it. The result was published in Russian only (see [1]).

Now I would like to present an extension of this axiomatization onto the first-order level.

For simplicity, I address the version of FOL that operates with relation and individual constants only (without identity). The only quantifier added is $\forall$ ($\exists$ is to be defined in the standard way). The truth-values truth, truth-value gap and falsehood are represented with 0, 1 and 2, respectively. The notion of signature (symbol set, set of constants) is defined in the usual way. A model of signature Σ is an ordered pair $M = \langle A, \alpha \rangle$ where A is a non-empty set and α a function mapping each n-placed predicate constant P to a partition $\langle P^{\alpha 0}, P^{\alpha 1}, P^{\alpha 2} \rangle$ of the set A^n and each individual constant c to an element of A. We accept $v_0, \ldots, v_n, \ldots$, as the set of individual variables and define their evaluation in the standard way.

A pair $J = \langle M, \beta \rangle$ where M is a model of signature Σ and β a variable evaluation is an interpretation. The meaning $\mathrm{mean}(\_, J)$ of terms and formulas under a given interpretation $J = \langle M, \beta \rangle$ of the corresponding signature is defined as follows:

(1) $\mathrm{mean}(c, J) = \alpha(c)$
(2) $\mathrm{mean}(v_i, J) = \beta(v_i)$
(3) $\mathrm{mean}(P(t_1, \ldots, t_n), J) = i$ iff $\langle \mathrm{mean}(t_1, J), \ldots, \mathrm{mean}(t_n, J) \rangle \in P^{\alpha i}$
(4) $\mathrm{mean}((\Phi \to \Psi), J) = 1$, if $\mathrm{mean}(\Phi, J) = 2$
(5) $\mathrm{mean}((\Phi \to \Psi), J) = \mathrm{mean}(\Psi, J)$, if $\mathrm{mean}(\Phi, J) < 2$
(6) $\mathrm{mean}(\neg\Phi, J) = (2 - \mathrm{mean}(\Phi, J))$
(7) $\mathrm{mean}(L\Phi, J) = 0$ iff $\mathrm{mean}(\Phi, J) = 0$, otherwise $\mathrm{mean}(L\Phi, J) = 2$
(8) $\mathrm{mean}(\forall v_i \Phi, J) = \max(\{\mathrm{mean}(\Phi, J') | J' = \langle M, \beta' \rangle$ and $\beta'(v_j) = \beta(v_j)$ for all $j \neq i\})$

Notions of satisfiability, validity, semantic consequence and inference are then defined in the usual way. Let $\Phi[t/v_i]$ be the result of replacement of the variable v_i with the term t in the formula Φ.

Consider the following Hilbert-style axiomatic system.

Axiom schemes:

    A1. $(\Phi \to (\Psi \to \Phi))$
    A2. $((\Phi \to (\Psi \to \Xi)) \to ((\Phi \to \Psi) \to (\Phi \to \Xi)))$
    A3. $(\neg(\Phi \to \Psi) \to (\Phi \to \neg\Psi))$
    A4. $((\neg\Phi \to \neg\neg\Phi) \to \Phi)$
    A5. $((L\Phi \to L\neg\Psi) \to \neg(\Psi \to \Phi))$
    A6. $((\neg\Phi \to \neg\Psi) \to (L\Psi \to L\neg\neg\Phi))$
    A7. $(\forall v_i \Phi \to \Phi[t/v_i])$
    A8. $(\neg\Phi[t/v_i] \to \neg\forall v_i \Phi)$
    A9. $(\forall v_i(\Phi \to \Psi) \to (\Phi \to \forall v_i \Psi))$ where v_i is not free in Φ
    A10. $(\forall v_i(\neg\Phi \to \Psi) \to (\neg\forall v_i \Phi \to \Psi))$ where v_i is not free in Ψ.

Inference rules:

    R1. $\Phi, (\Phi \to \Psi) \Rightarrow \Psi$
    R2. $\Phi \Rightarrow \forall v_i \Phi$

Theorem 1. A formula Φ of signature Σ is valid iff it is derivable from signature Σ substitution cases of **A1–A10** by **R1–R2**.

Proof. The 'if' part of the theorem follows from the fact that all Σ substitution cases of **A1–A10** are valid and **R1–R2** do preserve validity. The fact is easily verified on the basis of the definitions given above.

The 'only if' part can be derived from the lemma stating that every set of Σ formulas that is consistent with respect to the considered axiomatic system has a superset that has a model of a signature $\Sigma' \supseteq \Sigma$. This lemma can be proved in a way similar to that of analogous proofs in classical FOL. I just sketch some crucial steps here.

Consider a consistent set X of Σ. We substitute each variable that occurs free in some formulas of X with a new unique individual constant. This operation may result in an extension of Σ with some new constants and we name the least required extension Σ_0, and we name X' the set of statements (i.e. formulas without free variables) obtained from X by the described substitution. Of course, X' is consistent iff X is consistent. Call a set Y of Σ complete in Σ iff for any statement Φ of Σ either $\Phi \in Y$ or $L\neg\Phi \in Y$ holds. By Zorn's lemma, there is a consistent and complete in Σ_0 set X_0 such that $X \subseteq X_0$. Further, we construct a sequence $X_0, \ldots, X_n, \ldots$, and their respective signatures $\Sigma_0, \ldots, \Sigma_n, \ldots$, by the following induction. To obtain X_{i+1} on the basis of X_i we add a new unique constant c_Φ to Σ_i for each $\Phi \in X_i$ and we define Σ_{i+1} as the result of extending Σ_i with all the needed constants. Then for any statement $\neg\forall v_i\Phi \in X_i$ we add to X_i the new formula $\neg\Phi[c_\Phi/v_i]$. Thus we construct the set X_i' and we set X_{i+1} to be any complete in Σ_{i+1} superset of X_i'. Finally, we set Σ_ω to be the union of Σ_i for all natural i and X_ω to be the union of X_i for all natural i.

Now we state that X_ω has the following properties for arbitrary statements Φ, Ψ of Σ_ω:

P1 It is consistent

P2 Either $\Phi \in X_\omega$ or $L\neg\Phi \in X_\omega$

P3 Either $\neg\Phi \in X_\omega$ or $L\Phi \in X_\omega$

P4 At least one element of $\{\Phi, \neg\Phi\}$ is in X_ω

P5 $(\Phi \rightarrow \Psi) \in X_\omega$ iff $(\Phi \notin X_\omega$ or $\Psi \in X_\omega)$

P6 $\neg(\Phi \rightarrow \Psi) \in X_\omega$ iff $(\Phi \notin X_\omega$ or $\neg\Psi \in X_\omega)$

P7 $\neg L\Phi \in X_\omega$ iff $\neg\Phi \in X_\omega$

P8 $\neg\neg\Phi \in X_\omega$ iff $\Phi \in X_\omega$

P9 $\forall v_i\Phi \in X_\omega$ iff for all c from Σ_ω, $\Phi[c/v_i] \in X_\omega$

P10 $\neg\forall v_i\Phi \in X_\omega$ iff for some c from Σ_ω, $\neg\Phi[c/v_i] \in X_\omega$

P2–P4 imply that for an arbitrary statement Φ of Σ_ω there can be exactly three options: either $\Phi, L\Phi \in X_\omega$ and $\neg\Phi, L\neg\Phi \notin X_\omega$, or $\Phi, \neg\Phi \in X_\omega$ and $L\Phi, L\neg\Phi \notin X_\omega$, or $\neg\Phi, L\neg\Phi \in X_\omega$ and $\Phi, L\Phi \notin X_\omega$. Thus we define the required model $M = \langle A, \alpha \rangle$ for X_ω in the following way: A is the set of constants of Σ_ω and α a function such that for each n-placed predicate constant P and arbitrary n-tuple of constants $c_1, \ldots, c_n$, $\langle c_1, \ldots, c_n \rangle \in P^{\alpha 0}$ iff both $P(c_1, \ldots, c_n)$,and $LP(c_1, \ldots, c_n)$, are in X_ω, $\langle c_1, \ldots, c_n \rangle \in P^{\alpha 1}$ iff both $P(c_1, \ldots, c_n)$,and $\neg P(c_1, \ldots, c_n)$, are in X_ω, and $\langle c_1, \ldots, c_n \rangle \in P^{\alpha 2}$ iff both $\neg P(c_1, \ldots, c_n)$,and $L\neg P(c_1, \ldots, c_n)$, are in X_ω. It is easy to verify, by induction, that any statement of Σ_ω has a designated truth-value in M iff it is in X_ω. $\qquad \square$

Theorem 2. A1–A10 and **R1–R2** are pairwise independent.

Proof. Proofs of independence of **A2–A6** were given in [1]. To prove *independence of* **A1**, let $\forall v_i \Phi$ be equivalent to Φ and let $\{0, 1, 2, 3, 4\}$ be the set of truth values where the only non-designated value is 4. Let $(\Phi \rightarrow \Psi)$ take the value 0 iff the value of Ψ is 0 and the value of Φ is designated, take the value 1 iff the value of Ψ is in $\{1, 2\}$ or the value of Φ is 4, or the value of both Φ and Ψ is 3, take value 3 iff the value of Ψ is 3 and the value of Φ is in $\{0, 1\}$, and take the value 4 iff the value of Ψ is 4 and the value of Φ is designated, or the value of Ψ is 3 and the value of Φ is 2. Let $\neg \Phi$ take the value 0 iff the value of Φ is 4, take the value 4 iff the value of Φ is 0, and take the value 1 otherwise. Let $L\Phi$ take the value 0 iff the value of Φ is 0 and take the value 4 otherwise.

To prove *independence of* **A7**, let $\{0, 1, 2, 3\}$ be the set of truth values where the only designated values are 0 and 1. Let $(\Phi \rightarrow \Psi)$ take the value 0 iff the value of Ψ is 0 and the value of Φ is designated, take the value 1 iff the value of Ψ is 1 or the value of Φ is non-designated, and take value 2 iff the value of Ψ is non-designated and the value of Φ is designated. Let $\neg \Phi$ take the value 0 iff the value of Φ is non-designated, take the value 2 iff the value of Φ is 0, and take the value 1 otherwise. Let $L\Phi$ take the value 0 iff the value of Φ is 0 and to take the value 2 otherwise. For all v_i, let the value of $\forall v_i \Phi$ be 1 if the value of Φ is 3 and let the value of $\forall v_i \Phi$ be the value of Φ otherwise.

To prove *independence of* **A8**, define the values of $(\Phi \rightarrow \Psi)$, $\neg \Phi$, $L\Phi$ as in the main interpretation and let $\forall v_i \Phi$ take the value of $\neg L \neg \forall v_i \Phi$ in the main interpretation.

To prove *independence of* **A9**, let $\{0, 1, 2, 3\}$ be the set of truth values where the only non-designated value is 3. Let $(\Phi \rightarrow \Psi)$ take the value 0 iff the value of Ψ is 0 and the value of Φ is designated, take the value 1 iff the value of Ψ is in $\{1, 2\}$ or the value of Φ is 3, take value 3 iff the value of Ψ is 3 and the value of Φ is designated. Let $\neg \Phi$ take the value 0 iff the value of Φ is 3, take the value 3 iff the value of Φ is 0, and to take the value 1 otherwise. Let $L\Phi$ take the value 0 iff the value of Φ is 0 and take the value 3 otherwise. For all v_i, let the value of $\forall v_i \Phi$ be 1 if the value of Φ is 2 and let the value of $\forall v_i \Phi$ be the value of Φ otherwise.

To prove *independence of* **A10**, define the values of $(\Phi \rightarrow \Psi)$, $\neg\Phi$, $L\Phi$ as in the main interpretation and define $\forall v_i \Phi$ to take the value of $(L\neg\forall v_i \Phi \rightarrow \forall v_i \Phi)$ in the main interpretation.

Independence of **R1** and **R2** can be proved as in classical FOL. □

The distinction between designated values 0 and 1 can be taken up to simulate the difference between the program implementations that use hardware resources and those that do not. For example, implementing the command 'if P then A', a computer verifies the state P and then realizes A. If P holds but A is not realized then the computer fails to realize the command (value 2). If P holds and the computer realizes A, then the command is accomplished and some part of its hardware is used, if needed. Hence the command is accomplished with or without using hardware whenever A is accomplished with (value 0) or without (value 1) such a use. Finally, when P fails, the computer accomplishes the command without using any hardware (value 1).

References

[1] G. K. Olkhovikov. On a new three-valued paraconsistent logic. In *Logic of Law and Tolerance*, pages 96–113, Yekaterinburg, 2002. Ural State University Press. In Russian.

Received as reprint

A Comparison of Connexive Logics

Luis Estrada-González*
Instituto de Investigaciones Filosóficas
Universidad Nacional Autónoma de México, Mexico
loisayaxsegrob@gmail.com

Elisángela Ramírez-Cámara
Instituto de Investigaciones Filosóficas
Universidad Nacional Autónoma de México, Mexico
eliramirezc@gmail.com

Abstract

The most widespread criterion for the admission of a logic into the connexive family is the satisfaction of the pairs of formulas known as Aristotle's and Boethius' theses, along with the non-symmetry of implication. In this paper, we discuss whether this is enough to characterize a connexive logic or if more can be said about the issue. Our strategy is the following: first, we introduce a logic that has origins and motivations that have little to do with connexive logic. We then present a list of additional criteria found scattered throughout the literature on connexivity and propose to use this list to compare this logic and some of the well-known non-bivalent truth-functional connexive logics. This comparison gives us several interesting results: when every condition is given the same weight, the introduced logic can score as high as some of the well-known systems. Furthermore, a connection between the satisfaction of the most conditions and the loss of intuitiveness or an increase in the complexity of certain structural properties of the system seems to arise. We take these results to motivate the more general open problem of finding the most adequate way of judging systems of connexive logic.

We are extremely grateful to the two anonymous referees for their very helpful comments, and to the Editors of this special volume, especially Thomas Macaulay Ferguson, for their careful handling of the paper. Special thanks are deserved for the members of the FiCiForTes Seminar at UNAM.

*This paper has been written under the support from the PAPIIT project IA401015 "Tras las consecuencias. Una visión universalista de la lógica (I)", as well as from the CONACyT project CCB 2011 166502 "Aspectos filosóficos de la modalidad".

1 Introduction

This paper is part of a larger project aimed at investigating (1) what the necessary and sufficient conditions a logic should meet in order to belong to the connexive family are and (2) whether there is a known connexive logic that can be said to stand as the best one. More precisely, in this paper we collect and present what we take to be the most common desiderata to be met by any connexive logic. On that basis, we compare some of the most well-known non-bivalent truth-functional connexive logics. To make things more interesting, we also present MRS^P, a logic first introduced in [9], and use it as a sort of trial balloon for our list of criteria. Our motivations for this are the following: first, MRS^P arises from a technique different from the ones commonly used to get connexive logics—the algebraic, the ternary frames, the constructive, and the consequential ones. Second, both the technique and the resulting logic have a rationale independent from that of connexive logics, that is, they are not explicitly designed to validate Aristotle's Theses and related principles, but rather depend on general considerations about the proper valuation for conditionals with false antecedent. We believe that, with some argumentation, MRS^P could be shown to score high (as the best, even) among the connexive logics here considered. Nonetheless, we are not going to defend this claim, but rather encourage further investigation on the comparison of connexive logics.

The plan of the paper is as follows. Section 2 begins with a precise definition of the technique of weakening semantics. For simplicity, we restrict ourselves to the zero-order case. After that, we present the special weakening of the usual two-valued semantics for classical (zero-order) logic which characterizes the logic MRS^P. In Section 3, we collect and present what we take to be the most common desiderata to be met by a connexive logic. Finally, in Section 4 these desiderata become the starting point for a comparison of some of the most well-known non-bivalent truth-functional connexive logics. Here we include MRS^P as a test case, with the result that it scores rather high as a connexive logic if certain connexivist principles are loosened a bit. We take this to suggest that more remains to be done in order to determine the best way to compare any set of connexive logics.

2 The Logic MRS^P

The notion of *weakened semantics* was superficially introduced by that name in [9]. We will now offer a more formal presentation.

Let L be a formal language, V a collection of truth values partially ordered by

$\leq$ [1] (with at least two non-empty subcollections D^+ and D^- of *designated* and *anti-designated* values, respectively), and SEM a collection of interpretation functions of the form $\sigma : L \times I \longrightarrow V$, where I is a collection of indexes of evaluation with a certain relation R among them. A *weakened semantics of L based on SEM, SEM^W*, is a collection of interpretations $\sigma^W : L \times I^* \longrightarrow V^*$ such that:

(WS1) $V \subseteq V^*$, $D^+ \subseteq D^{+*}$, $D^- \subseteq D^{-*}$ (with D^{+*} and D^{-*} as non-empty subcollections of V^*), $I \subseteq I^*$ (and $R \subseteq R^*$) and there is at least one v^* in V^* and a v in V such that $v^* < v$ in V^*;

(WS2) for some $A_1, \dots, A_n$ and n-ary connective k, there are a σ and a σ^W such that either $\sigma(k(A_1, \dots, A_n)) = v$ and $\sigma^W(k(A_1, \dots, A_n)) = v^*$, or, for each A_i, $\sigma^W(k(A_1, \dots, A_n)) < \sigma^W(A_i)$;

(WS3) if there are two n-ary connectives c, k such that $c(A_1, \dots, A_n)$ and $k(A_1, \dots, A_n)$ are L-formulas, and there is a valuation σ such that $\sigma(c(A_1, \dots, A_n)) \neq \sigma(k(A_1, \dots, A_n))$, then there is a σ^W such that $\sigma^W(c(A_1, \dots, A_n)) \neq \sigma^W(k(A_1, \dots, A_n))$ too; and

(WS4) the notion of logical consequence associated to SEM^W must be at most as strong as the notion of logical consequence associated to SEM.

Proper weakening can be attained by adding the following condition:

(PW) There are no A in L, v^* in V^* and v in V such that $v < v^*$ in V^*, $\sigma(A) = v$ and $\sigma^W(A) = v^*$.

The informal reading of the above is this. The weak semantics must have enough resources (WS1) to ensure that some formulas get a value lesser than the one they would obtain under the original interpretation (WS2). Nonetheless, the weakening of the semantics must be such that different connectives remain distinguishable after the weakening (WS3) and the notion of logical consequence associated to SEM^W must be at most equivalent to the notion of logical consequence associated to SEM (WS4). Finally, the intuitive reading of (PW) is that even if weakening allows for making formulas less true than they were under the original SEM, a proper weakening should not allow interpretations that make formulas "truer" than they were under the original SEM.

A notion of *strengthening semantics* can be defined too. It is like weakening semantics, except that certain orderings in (WS1), (WS2) and (WS4) are inverted. Thus,

---

[1]Every partial order $\leq$ induces a *strict order*, defined as $x < y = (x \leq y$ and $x \neq y)$.

(SS1) $V \subseteq V^*$, $D^+ \subseteq D^{+^*}$, $D^- \subseteq D^{-^*}$ (with D^{+^*} and D^{-^*} as non-empty subcollections of V^*), $I \subseteq I^*$ (and $R \subseteq R^*$), and there is at least one v^* in V^* and a v in V such that $v < v^*$ in V^*;

(SS2) for some $A_1, \ldots, A_n$ and n-ary connective k, there are σ and σ^S such that either $\sigma(k(A_1, \ldots, A_n)) = v$ and $\sigma^S(k(A_1, \ldots, A_n)) = v^*$, or, for each A_i, $\sigma^S(k(A_1, \ldots, A_n)) < \sigma^S(A_i)$;

(SS3) if there are two n-ary connectives c, k such that $c(A_1, \ldots, A_n)$ and $k(A_1, \ldots, A_n)$ are L-formulas, and there is a valuation σ such that $\sigma(c(A_1, \ldots, A_n)) \neq \sigma(k(A_1, \ldots, A_n))$, then there is a σ^S such that $\sigma^S(c(A_1, \ldots, A_n)) \neq \sigma^S(k(A_1, \ldots, A_n))$ too; and

(SS4) the notion of logical consequence associated to SEM^S must be at least as strong as the notion of logical consequence associated to SEM.

Note that (WS3) goes unchanged and (SS2) is typographically identical to (WS2), but its ordering is dependent on the first clause. As expected, *proper strengthening* is also definable:

(PS) There are no v^* in V^* and v in V such that $v^* < v$ in V^*, $\sigma(A) = v$ and $\sigma^S(A) = v^*$.

MRS^P was defined as a logic obtained by weakening the usual two-valued semantics for classical zero-order logic as follows:

If in the classical case $V = \{\bot, \top\}$ with $\bot < \top$, let $V^* = \{\bot, *, \top\}$, with $\bot < * < \top$ and $D^+ = D^{+^*} = \{\top\}$. (Given that in the classical case there is only one index of evaluation with the identity relation as R, it can be left implicit.) With the only exception of the conditional, the satisfiability conditions for connectives are the usual ones:

- $\sigma^{W^*}(\neg A) = \begin{cases} \bot & \text{if } \sigma^{W^*}(A) = \top \\ \top & \text{otherwise} \end{cases}$ f.

- $\sigma^{W^*}(A \wedge B) = \inf(\sigma^{W^*}(A), \sigma^{W^*}(B))$.

- $\sigma^{W^*}(A \vee B) = \sup(\sigma^{W^*}(A), \sigma^{W^*}(B))$.

- $\sigma^{W^*}(A \equiv B) = \begin{cases} \top & \text{if } \sigma^{W^*}(A) = \sigma^{W^*}(B) \\ \inf(\sigma^{W^*}(A), \sigma^{W^*}(B)) & \text{otherwise} \end{cases}$.

$$\bullet\ \sigma^{W^*}(A \supset B) = \begin{cases} \sigma^{W^*}(B) & \text{if } \sigma^{W^*}(A) = \top \\ * & \text{otherwise} \end{cases}.$$

Or, using truth tables:

A	B	$\neg A$	$A \wedge B$	$A \vee B$	$A \supset B$	$A \equiv B$
$\top$	$\top$	$\bot$	$\top$	$\top$	$\top$	$\top$
$\top$	$*$	$\bot$	$*$	$\top$	$*$	$*$
$\top$	$\bot$	$\bot$	$\bot$	$\top$	$\bot$	$\bot$
$*$	$\top$	$\top$	$*$	$\top$	$*$	$*$
$*$	$*$	$\top$	$*$	$*$	$*$	$\top$
$*$	$\bot$	$\top$	$\bot$	$*$	$*$	$\bot$
$\bot$	$\top$	$\top$	$\bot$	$\top$	$*$	$\bot$
$\bot$	$*$	$\top$	$\bot$	$*$	$*$	$\bot$
$\bot$	$\bot$	$\top$	$\bot$	$\bot$	$*$	$\top$

It is easy to verify that this semantics satisfies the conditions to be a weakening of the usual two-valued semantics for classical logic.[2] One finds that $* < \top$ from looking at the proposed V^*, and clearly $D^+ = D^{+*}$, so the first condition is met. From the weakened satisfiability conditions for the conditional, one can see that $\sigma^{W^*}(A \supset B) = *$ whenever $\sigma^{W^*}(A) = \bot$; in the classical case, the value of the conditional is $\top$ when the antecedent is $\bot$. The truth table above shows clearly that all of $(A \wedge B)$, $(A \vee B)$, $(A \supset B)$, $(A \equiv B)$ remain in general non-equivalent to each other. The notion of logical consequence is the Tarskian notion of consequence usually associated with classical logic, which is reflexive, transitive and monotonic. The reader can easily verify that, moreover, this weakening is proper.

The logic MRS^P was introduced in [9] as a logic to validate more relations in the square of opposition, especially subalternation, hence the name.[3] The proposed solution was to get a logic in which the inference from a conditional $A \supset B$ to the conjunction $A \wedge B$ never leads from truth to falsity. Besides subalternation and some other relations of opposition, Aristotle's Thesis together with other connexivist principles were validated.

Thus, the connexivist ideas on implication could serve as a conceptual foundation for MRS^P as a whole. Even so, a motivation for assigning the intermediate value

---

[2]Note that even if the biconditional cannot be defined in the usual way, i.e. as $(A \supset B) \wedge (B \supset A)$, its satisfiability conditions are the usual ones, and given that the satisfiability conditions of the conjunction are also the standard ones, what should be blamed—if at all—for the failure of the equivalence are either the structure of V^* or the satisfiability conditions for the conditional.

[3]In [9] the superscript 'P' was used to indicate, mistakenly, that negation is Post's cyclic negation. Despite this, we are going to keep the name.

* to a conditional when its antecedent is $\perp$ is needed to make the logic something more than a nearly *ad hoc* way of validating Aristotelian inferences or connexivist principles. On that note, we would like to point out that the satisfiability conditions of the conditional in MRS^P coincide with those of Blamey's *transplication* (see [5]). This means that Blamey's motivations for such a connective—for example, Strawsonian presupposition or conditional assertion—can be invoked (more about this can be found in [4]). Furthermore, Edgington-like arguments against the usual satisfiability conditions for conditionals with a false antecedent could be given (cf. [7], [8]; see also [14]), with the caution that her arguments against the other clauses should be either blocked or ignored!

Given that MRS^P validates Aristotle's Theses—something considered a distinguishing mark of a connexive logic—without being explicitly designed for this purpose, one may wonder whether it has some other properties typical of connexive logics. In order to find out how connexive MRS^P really is, one may compare it with other more famous logics in that family. In the next two sections, we address such a task. We begin by isolating what seem to be some of the typical properties of a connexive logic in the one that follows.

3 Desiderata for Connexivity and Kinds of Connexive Logics

Connexive logics codify certain ideas about the connection or coherence between the relata of an implication. Thus far there is no comprehensive, systematic study on what the properties of a connexive logic should be. This section aims to be a small step in that direction. Here we have identified, and labeled when necessary, some of the desiderata for a connexive logic found scattered through the literature, but particularly in the surveys [13] and [18].

In accordance with the mentioned literature, we take the following to be the minimal requirements for a connexive logic:

$\Vdash \neg(A \supset \neg A)$ (Aristotle's Thesis)
$\Vdash \neg(\neg A \supset A)$ (Variant of Aristotle's Thesis)
$\Vdash (A \supset B) \supset \neg(A \supset \neg B)$ (Boethius' Thesis)
$\Vdash (A \supset \neg B) \supset \neg(A \supset B)$ (Variant of Boethius' Thesis)[4]
$\nVdash (A \supset B) \supset (B \supset A)$ (Non-Symmetry of Implication)

---

[4]An anonymous reviewer pointed out that we were not considering the satisfaction of the converses of both Boethius' Thesis and its variant as a condition for minimality. We think that the satisfaction of these theses is perhaps better suited to be taken as an additional criterion when comparing connexive logics. However, we decided not to include them, as the theses are not validated

A *subminimal connexive logic* is a logic which satisfies at least some but not all of the above conditions. Specifically, a *subminimal connexive logic* is one which satisfies at least some of the positive conditions.

However, the story about connexivity does not end here. Some (sub)minimal connexive logics have been judged or evaluated, more or less consciously, by appealing to some other properties besides minimality (we will sketch how this has been done in the following section). Again, we extract these other desiderata from the available literature.

An *Abelardian logic* satisfies either of the following requirements:

$$\Vdash \neg((A \supset B) \wedge (\neg A \supset B)) \qquad \text{(Aristotle's Second Thesis)}$$
$$\Vdash \neg((A \supset B) \wedge (A \supset \neg B)) \qquad \text{(Abelard's Principle)}[5]$$

A logic is *anti-paradox* if it satisfies the following requisites:

$$\not\Vdash A \supset (B \supset A) \qquad \text{(Positive Paradox of Implication)}$$
$$\not\Vdash A \supset (\neg A \supset B) \qquad \text{(Negative Paradox of Implication)}$$
$$\not\Vdash A \supset (B \supset C) \qquad \text{where } A \text{ is a contingent truth and } (B \supset C)$$
$$\text{a logical truth (Paradox of Necessity)}$$

A logic is *simplificative* if it meets the following conditions:

$$\Vdash (A \wedge B) \supset A \qquad \text{(Simplification (a))}$$
$$\Vdash (A \wedge B) \supset B \qquad \text{(Simplification (b))}$$

A *conjunction-idempotent logic* is a logic in which both of the following conditions hold:

$$\Vdash (A \wedge A) \supset A \qquad \text{(Idempotence (a))}$$
$$\Vdash A \supset (A \wedge A) \qquad \text{(Idempotence (b))}$$

A *weakly consistent logic* is a logic in which there is no formula A such that it and $\neg A$ are both theorems; it is *strongly inconsistent* otherwise. A *strongly consistent logic* is a logic in which there is no formula A such that it and $\neg A$ are both satisfiable; it is *weakly inconsistent* otherwise.

Furthermore, a logic is *Kapsner-strong* if the two following conditions are met:

$A \supset \neg A$ is unsatisfiable

$A \supset B$ and $A \supset \neg B$ are not simultaneously satisfiable.

These two conditions were introduced in [10] to characterize *strongly connexive logics*, that is, minimal connexive logics which are also Kapsner-strong. We emphasize

---

by many connexive logics and seem to have counterexamples even by connexivist lights. See [13, 446] for further discussion.

[5]Also known as 'Strawson's Thesis'.

the distinction between a logic being Kapsner-strong and being strongly connexive to allow for the possibility of a logic being both subminimal and Kapsner-strong.

Finally, a logic is *totally connexive* if it is of all the other kinds above (only satisfying, of course, the appropriate versions of (in)consistency, as no logic can satisfy all of them). Clearly, there are some obstacles to get a totally connexive logic. For example, it is well-known that a consistent connexive logic cannot be simplificative if it also satisfies *Contraposition* for the conditional and *Transitivity* of logical consequence. The same goes for a logic that satisfies *Detachment* and validates $((A \supset B) \supset (B \supset C)) \supset (A \supset C)$. As a result, an open problem is to find out whether there are totally connexive logics and what is the minimal one.

In non-bivalent contexts, where the *Deduction (meta)theorem* might not be available, it is extremely useful, and even conceptually mandatory, to distinguish between $\Vdash A \supset B$ and $A \Vdash B$. For the purposes of this paper, it will be useful to consider a specific variety of connexive logics. Let K be a set of conditions for a connexive logic of the form $\Vdash A_1 \supset B_1; \ldots; \Vdash A_n \supset B_n$. A logic will be called *inferentially K connexive logic* if it satisfies $A_1 \Vdash B_1; \ldots; A_n \Vdash B_n$.

4 Comparison with CC1, M3V and CN

The above list of desiderata for a connexive logic is not meant to be exhaustive, and is far from being a set of necessary and sufficient conditions for connexivity. It does, however, give us some working data to compare a very specific selection of connexive logics. That is, it allows us to compare some of the most well-known non-bivalent truth-functional connexive logics: Angell's CC1, Mortensen's M3V and Cantwell's CN, to which we will add MRS^P. For simplicity and to avoid making the paper unnecessarily long, we will assign the same weight to each of the criteria. Certainly, there might be reasons to favor some of them above the others—for example, strong consistency and simplificativeness over Abelardianism, thus reviving the twelfth century crisis in logic (see [11]). Also for brevity we omit the proofs of the claims that a logic satisfies such and such properties. In most cases, the proofs are already well-known, not too difficult, and can be found somewhere else in the easily traceable literature on connexive logics. Nonetheless, we will invariably give the required counterexamples when we say that a logic fails to satisfy certain properties. These counterexamples may also be easy to find, well-known and traceable, but their inclusion makes for a smoother reading, as it makes the paper self-contained with respect to the inclusion of elements that facilitate the comparisons.

CC1 was introduced in [2] and is the first logic of the latest generation of studies on connexive implication, which began in the 1960s. It is characterized by the following truth tables with the values[6] $V_{CC1} = \{\top^+, \top^-, \bot^+, \bot^-\}$ ordered $\bot^- < \bot^+ < \top^- < \top^+$ and with $D^+ = \{\top^+, \top^-\}$:

A	B	$\neg A$	$A \land B$	$A \supset B$
$\top^+$	$\top^+$	$\bot^-$	$\top^+$	$\top^+$
$\top^+$	$\top^-$	$\bot^-$	$\top^-$	$\bot^-$
$\top^+$	$\bot^+$	$\bot^-$	$\bot^+$	$\bot^+$
$\top^+$	$\bot^-$	$\bot^-$	$\bot^-$	$\bot^-$
$\top^-$	$\top^+$	$\bot^+$	$\top^-$	$\bot^-$
$\top^-$	$\top^-$	$\bot^+$	$\top^+$	$\top^+$
$\top^-$	$\bot^+$	$\bot^+$	$\bot^-$	$\bot^-$
$\top^-$	$\bot^-$	$\bot^+$	$\bot^+$	$\bot^+$
$\bot^+$	$\top^+$	$\top^-$	$\bot^+$	$\top^+$
$\bot^+$	$\top^-$	$\top^-$	$\bot^-$	$\bot^-$
$\bot^+$	$\bot^+$	$\top^-$	$\bot^+$	$\top^+$
$\bot^+$	$\bot^-$	$\top^-$	$\bot^-$	$\bot^-$
$\bot^-$	$\top^+$	$\top^+$	$\bot^-$	$\bot^-$
$\bot^-$	$\top^-$	$\top^+$	$\bot^+$	$\top^+$
$\bot^-$	$\bot^+$	$\top^+$	$\bot^-$	$\bot^-$
$\bot^-$	$\bot^-$	$\top^+$	$\bot^+$	$\top^+$

To avoid making the truth table unnecessarily confusing, we will just point out that disjunction and the biconditional for CC1 can be defined in the usual ways—as $\neg(\neg A \land \neg B)$ and $(A \supset B) \land (B \supset A)$, respectively. Before discussing which conditions are met by CC1, we would like to note in passing that its matrices could be seen as the result of both a weakening and a strengthening of any n-valued truth-functional logic, with $n < 4$. For example, if $\sigma(A) = \sigma(B) = \bot^-$, $\sigma(A \land B)$ is not $\bot^-$ as one would expect, but $\bot^+$; since $\bot^- < \bot^+$, this particular σ can be regarded as a strengthening of the satisfiability conditions for $\land$ of most $n(< 4)$-valued truth-functional logics. Another example of strengthening for the conjunction is given by $\sigma(A) = \sigma(B) = \top^-$, which yields $\sigma(A \land B) = \top^+$. On the other hand, $\sigma(A) = \top^+$ and $\sigma(B) = \top^-$, as well as $\sigma(A) = \top^-$ and $\sigma(B) = \top^+$, would give an example of weakened $\sigma(A \supset B)$. The reader is kindly asked to check whether something similar happens with the other logics presented in this section.

Among the desiderata for a connexive logic, CC1 is minimal, (strongly) consistent, Abelardian and Kapsner-strong. As is well-known since the works by Mc-

---

[6]Since we are mostly going to discuss the structural properties of truth values, no deep consequences should be extracted from typographical identity; in particular, that two logics share, say, the symbol '⊥' for their least value does not imply that the intended interpretation for such a symbols is the same in those two logics.

Call ([12]) and Routley and Montgomery ([16]), it fails to be simplificative[7] and idempotent[8]. It also fails to meet anti-paradoxicality, for any formula of the form $(A \wedge A) \supset (B \supset B)$ is a theorem of CC1.

The logic M3V was introduced, although not with that name, in [15] (the name was given in [13], presumably to mean "Mortensen's 3-valued connexive logic"). Its biconditional is also defined as usual, that is, as $(A \supset B) \wedge (B \supset A)$. The following truth tables, with $V_{M3V} = \{\top, \top^*, \bot\}$ ordered $\bot < \top^* < \top$, and with $D^+ = \{\top, \top^*\}$, characterize M3V:

A	B	$\neg A$	$A \wedge B$	$A \vee B$	$A \supset B$
$\top$	$\top$	$\bot$	$\top$	$\top$	$\top^*$
$\top$	$\top^*$		$\top^*$	$\top$	$\bot$
$\top$	$\bot$		$\bot$	$\top$	$\bot$
$\top^*$	$\top$	$\top^*$	$\top^*$	$\top$	$\top^*$
$\top^*$	$\top^*$		$\top^*$	$\top^*$	$\top^*$
$\top^*$	$\bot$		$\bot$	$\top^*$	$\bot$
$\bot$	$\top$	$\top$	$\bot$	$\top$	$\top^*$
$\bot$	$\top^*$		$\bot$	$\top^*$	$\top^*$
$\bot$	$\bot$		$\bot$	$\bot$	$\top^*$

M3V is minimal, anti-paradox[9], Abelardian, simplificative and idempotent. However, it is strongly inconsistent[10] and fails to be Kapsner-strong[11].

CN is a logic for "conditional negation" studied in [6]. As with the other logics, its biconditional can be defined as $(A \supset B) \wedge (B \supset A)$. After considering that for CN the following apply: $V_{CN} = \{\top, -, \bot\}$ ordered $\bot < - < \top$, with $D^+ = \{\top, -\}$; one can build the following truth tables for its other connectives:

---

[7]For a counterexample to $(A \wedge B) \supset A$, take $\sigma(A) = \top^+$ and $\sigma(B) = \top^-$; for a counterexample to $(A \wedge B) \supset B$, take $\sigma(A) = \top^-$ and $\sigma(B) = \top^+$.

[8]For a counterexample to $(A \wedge A) \supset A$ take $\sigma(A) = \top^-$; for a counterexample to $A \supset (A \wedge A)$ take $\sigma(A) = \bot^-$.

[9]Actually, Mortensen's satisfiability conditions for the conditional are structurally the same as the ones used by Anderson and Belnap in [1] to block the paradox of necessity.

[10]Any formulas of the form $A \supset A$ and $\neg(A \supset A)$ are theorems of M3V.

[11]A formula of the form $A \supset \neg A$ is satisfiable when $\sigma(A) = \bot$; moreover, $A \supset B$ and $A \supset \neg B$ are simultaneously satisfiable with $\sigma(A) = \sigma(B) = \bot$.

A	B	$\neg A$	$A \wedge B$	$A \vee B$	$A \supset B$
⊤	⊤	⊥	⊤	⊤	⊤
⊤	−		−	⊤	−
⊤	⊥		⊥	⊤	⊥
−	⊤	−	−	⊤	⊤
−	−		−	−	−
−	⊥		⊥	−	⊥
⊥	⊤	⊤	⊥	⊤	−
⊥	−		⊥	−	−
⊥	⊥		⊥	⊥	−

CN is a minimal connexive logic; moreover, it is Abelardian, simplificative and idempotent. However, it fails to be strongly consistent, does not avoid all the paradoxes of implication[12] and is not Kapsner-strong[13].

What about MRS^P? It is only subminimal, since it does not validate Boethius' Theses[14]; it is Abelardian, anti-paradox, strongly consistent and Kapsner-strong. However, it is neither simplificative nor idempotent.[15]

Although the above comparison looks rather unfavorable for MRS^P, there is more to be said in its favor. For instance, one can wonder what resources for achieving the properties each logic needs. Accordingly, one can consider:

The number of designated values Roughly, the idea is that something can be said in favor of the thought that, whenever possible, one should obtain theoremhood and validity only with truth, not with truth and something lesser than it. Although it is not logically necessary that this be the case, several logics take the distinction between truth and any other truth value as fundamental. Of course, this could be generalized to a bipartition between designated and anti-designated values. Such a bipartition comes into play in debates regarding, for example, pluralism about truth predicates (See [3] and [17] for an introduction to both sides of the debate). We think that an analogy can be traced between this debate and what we are trying to establish with this condition. However, it is important to note that the matter is far from settled as the analogy can be drawn towards both sides of the debate.

The intuitiveness of the satisfiability conditions The idea is, broadly put, that there seems to be no disagreement between the valuations of the logics considered and

---

[12]It validates both the Positive Paradox and the Paradox of Necessity by validating $A \supset (A \supset A)$, for example.

[13]$A \supset \neg A$ is satisfiable with $\sigma(A) = -$ and even with $\sigma(A) = \bot$. This assignment suffices to also satisfy $A \supset B$ and $A \supset \neg B$ simultaneously.

[14]For an easy counterexample, consider $\sigma(A) = \sigma(B) = *$.

[15]For a counterexample to these latter properties, take $\sigma(A) = *$.

the ones for the classical case when they are restricted to the greatest and the least values in the semantics, and this should generalize as much as possible for the case of designated and antidesignated values. For example, there is no disagreement between the evaluation of $A \supset B$ offered by the logics considered and the classical case when both $\sigma(A)$ and $\sigma(B)$ are the greatest value—in every case, the whole conditional has the greatest value. Similarly, when $\sigma(A)$ is the greatest value and $\sigma(B)$ the least one—the value of the whole conditional is the least one. In this sense, it would be reasonable to expect that when both $\sigma(A)$ and $\sigma(B)$ are designated values, the whole conditional is also interpreted as a designated value; or that when $\sigma(A)$ is designated and $\sigma(B)$ is antidesignated, the whole conditional is antidesignated too.

If these two criteria are taken into consideration, then MRS^P achieves at least as much as CN—the satisfaction of four properties—with fewer logical resources, in the sense that it only requires one designated value. It should be noted, however, that this only works as an advantage if it is granted that both logics are on equal footing as to the intuitiveness of their respective satisfiability conditions. Moreover, if one finds consistency desirable, MRS^P gains an advantage over those of the examined proposals that have more designated values than antidesignated ones. In particular, MRS^P's strong consistency would make it score better than CN, as the latter is only weakly consistent.

The concession about the intuitiveness of their satisfiability conditions cannot be as easily made for either CC1 or M3V. While it is true that they meet the most properties out of all the proposals considered, they only do so at what seems to be a very high price to pay. In the case of CC1, the truth tables are very anomalous for a linear ordering, and seem more like "a formal tool with little explanatory power", as Wansing puts it (cf. [18]). This results in several odd interpretations: The conjunction of two non-greatest values yields a greatest value and the conjunction of two non-least values yields a least value, as can be seen in the sixth and tenth lines of the truth table. Furthermore, conditionals with designated values as both the antecedent and consequent yield anti-designated values. M3V is in no better position: Even for most contradictions' friends, these are very rare, and certainly very few of these people would propose logics with pairs of contradictory theorems. Moreover, the satisfiability conditions for M3V imply that a conditional might have an anti-designated value even if both the antecedent and the consequent are designated values.

As a final point of comparison, given that no logic considered here is total, we will now return to the notion of *inferentially K connexive logic*, taking K to be the set of the inferential versions of the desiderata listed above, to check whether any of the logics above can count as *inferentially total*. Under this examination, MRS^P is inferentially minimal, inferentially simplificative and inferentially idempotent. This, together with the properties already satisfied, makes MRS^P an inferentially total connexive logic. As a result, MRS^P gains some advantage over M3V and CN, as neither consistency nor Kapsner-strength have the required format for their inferential versions to be considered. The requisite of anti-paradoxicality does have it, but this does not make any difference for the comparison with CN, as the inferences from A to $B \supset A$ and from A to $A \supset A$ would be valid.

Things are a bit different for CC1. It is inferentially simplificative and inferentially idempotent, which means it fulfills the only two properties it needed to be inferentially total, against the three needed by MRS^P. However, it should be recalled that CC1 can achieve this only because of the oddness of its four-valued matrices. As CC1 is not only inferentially total, but also needs to meet fewer properties to be so, one might find it superior to MRS^P. However, if one considers the odd matrices used by CC1 and recalls the inclusion of considerations about the simplicity of the structure of designated values and the intuitiveness of the satisfiability conditions in the comparison, then one might take this to fairly tip the scales in favor of MRS^P.

But the connexive superiority of MRS^P is far from being established. It is only subminimal, and thus one could reasonably argue that MRS^P had already lost very early in the race. Another point against MRS^P is that the notion of inferential connexivity is anything but connexive: An important part of the connexive effort was to reflect in the object language some metatheoretical ideas about implication, and inferential connexivity has nothing to do with these ideas. Saying that MRS^P satisfies Boethius inferentially could be regarded as artificial as saying that classical logic is connexive because it satisfies Aristotle's Thesis in the form $A \nvdash \neg A$.

The only moral that can be drawn at this point is that it is extremely difficult to rank the logics studied here, as each of them exhibits very different combinations of virtues and problems. Were MRS^P at least minimal, a good case for it could be made. Regardless, it should be stressed that the other logics achieve minimality at the expense of intuitiveness or increasing the number of truth values. Our hope is that, rather than to achieve a consensus about which logic is the definitive winner, this discussion will serve to motivate a debate between the experts on connexive logic to rank the desiderata above, including intuitiveness and structural properties like the number of designated values or a certain degree of conservativeness over classical logic (for example, not departing radically from the classical valuations involving just the least and greatest elements). This debate, we believe, will lead to the development of better tools for logic choice or, at least, to a better mapping of the connexivist landscape.

5   Conclusions

In this paper we presented a roster of what we take to be the most common desiderata to be met by a connexive logic. Using the list and some additional criteria, we compared some well-known many-valued truth-functional connexive logics, namely Angell's CC1, Mortensen's M3V and Cantwell's CN. To these we have added MRS^P, a logic known to validate Aristotle's Theses, so one can legitimately ask what is its place in the universe of connexive logics. We showed that, with some pressure, one can conclude that MRS^P scores better than the other logics analyzed here.

Nevertheless, any case for favoring one of the logics above the others is far from conclusive. Thus, we take that the right conclusion of the analysis is that there is an important open problem of conceptual nature in the study of connexive logic, one that has to do with the investigation of necessary and sufficient conditions for connexivity, and with finding the most adequate criteria for the comparison of connexive logics that are not necessarily truth-functional, based on those conditions.

Along the way, we also encountered other open problems, more formal in nature, that require further attention even if the above problem resists solution. For example, there is the problem of knowing under what conditions a SEM which does not characterize a connexive logic can be (properly) weakened into a semantics SEM^W that does characterize one. An analogous problem can be formulated for (properly) strengthened semantics. Finally, another open problem is to know whether there are total connexive logics as have been defined above and, if there are, which is the minimal one.

References

[1] A.R. Anderson and N.D. Belnap, Jr. *Entailment: The Logic of Relevance and Necessity, Vol. I.* Princeton University Press, Princeton, 1975.

[2] R.B. Angell. A propositional logic with subjunctive conditionals. *Journal of Symbolic Logic*, 27(3):327–343, 1962.

[3] JC Beall. On mixed inferences and pluralism about truth predicates. *The Philosophical Quarterly*, 50(200):380–382, 2000.

[4] N.D. Belnap, Jr. Conditional assertion and restricted quantification. *Noûs*, 4(1):1–13, 1970.

[5] S. Blamey. Partial logic. In D. Gabbay and F. Guenthner, editors, *Handbook of Philosophical Logic*, volume III, pages 1–70. Kluwer, Dordrecht, 1986.

[6] J. Cantwell. The logic of conditional negation. *Notre Dame Journal of Formal Logic*, 49(3):245–260, 2008.

[7] D. Edgington. Do conditionals have truth-conditions? *Crítica. Revista Hispanoamericana de Filosofía*, 18(52):3–30, 1986.

[8] D. Edgington. On conditionals. *Mind*, 104(414):235–329, 1995.

[9] L. Estrada-González. Weakened semantics and the traditional square of opposition. *Logica Universalis*, 2(1):155–165, 2008.

[10] A. Kapsner. Strong connexivity. *Thought: A Journal of Philosophy*, 1(2):141–145, 2012.

[11] C. J. Martin. Embarrassing arguments and surprising conclusions in the development of theories of the conditional in the twelfth century. In J. Jolivet and A. de Libera, editors, *Gilbert de Poitiers et ses contemporains*, pages 377–401. Bibliopolis, Ed. di Filos. e Scienze, Naples, 1987.

[12] S. McCall. Connexive implication. *Journal of Symbolic Logic*, 31(3):415–433, 1966.

[13] S. McCall. A history of connexivity. In D. Gabbay and F. Guenthner, editors, *Handbook of the History of Logic*, volume 11, pages 415–449. Elsevier, Amsterdam, 2012.

[14] M. McDermott. On the truth conditions of certain 'if'-sentences. *Philosophical Review*, 105(1):1–37, 1996.

[15] C. Mortensen. Aristotle's thesis in consistent and inconsistent logics. *Studia Logica*, 43(1-2):107–116, 1984.

[16] R. Routley and H. Montgomery. On systems containing Aristotle's thesis. *Journal of Symbolic Logic*, 33(1):82–96, 1968.

[17] C. Tappolet. Truth pluralism and many-valued logics: A reply to Beall. *The Philosophical Quarterly*, 50(200):382–385, 2000.

[18] H. Wansing. Connexive logic. In E. N. Zalta, editor, *The Stanford Encyclopedia of Philosophy*. Fall 2014 edition, 2014.

Received October 2015

On Arithmetic Formulated Connexively

Thomas Macaulay Ferguson
Department of Philosophy
City University of New York Graduate Center, United States
tferguson@gradcenter.cuny.edu

Abstract

In this paper, we reflect on some themes related to the formulation of mathematics against the backdrop of a connexive logic. From a positive perspective, we will consider some remarks of the Kneales concerning Aristotle's position on connexive implication and suggest that common themes between the Kneales' Aristotle and the hyper-constructive arithmetic of David Nelson may provide a philosophical basis for connexive mathematics. We will also consider some historical points, including Łukasiewicz' argument that connexive principles may be refuted by appeal to number-theoretic intuitions. Finally, we will take more concrete steps towards the implementation of connexive mathematics by examining how weak subtheories of arithmetic fare when formulated in modest first-order extensions of three connexive logics: Richard Angell's PA1 and PA2 and Graham Priest's P_N. Unfortunately, we will observe that severe pathologies emerge when even extraordinarily weak subsystems of Peano arithmetic are evaluated in these logics, suggesting that Angell and Priest's systems constitute strained, if not unserviceable, bases for arithmetic.

1 The Allure of Connexive Mathematics

Frequently, non-classical logics are presented as formalizations of correct deductive reasoning and mathematical reasoning, as an *a priori* discipline, is uniquely sensitive to the adoption or rejection of logical principles. The analysis of how various mathematical theories fare under enriched or restrained theories of inference makes up one of the most salient applications of a non-classical theory of deduction. Historically, this is most evident in the case of intuitionism, insofar as the

I am very grateful for some helpful comments due to Maarten McKubre-Jordens and Can Başkent when the material was presented at the Workshop on Connexive Logic at the Fifth World Congress and School on Universal Logic. I also appreciate the very helpful comments of two anonymous referees.

intuitionistic standpoints with respect to deduction and mathematical practice are tightly bound together. But similar sentiments apply to mathematics investigated in other non-classical settings. For example, the analysis of mathematical principles by *substructural* logics has been particularly fruitful, including Robert Meyer's investigations into the relevant arithmetic $R^{\sharp}$,[1] John K. Slaney, Greg Restall, and Meyer's investigation into the linear arithmetic $LL^{\sharp}$ in [20], and Zach Weber's more recent investigations into set theory in weak relevant logics (*e.g.*, [24] and [25]). The concern of this work is to begin investigating the prospects for a *connexive* formulation of mathematics.

Connexive logics are deductive systems that contain one or more of the following as theorems:

$$
\begin{array}{ll}
\text{Aristotle's Thesis} & \sim(\varphi \to \sim\varphi) \\
\text{Boethius' Thesis} & (\varphi \to \psi) \to \sim(\varphi \to \sim\psi) \\
\text{Strawson's Thesis} & \sim((\varphi \to \psi) \wedge (\varphi \to \sim\psi))
\end{array}
$$

where "$\to$" represents a binary conditional connective, possibly interpreted as material implication (*e.g.*, in the classical case) or as an intensional entailment connective. The present examination of mathematics formulated with a connexive background logic is not unique, as there have been several forays into connexive mathematics in the modern era of connexive logic. Some of these investigations have been harmonious with reasonable and salient mathematical notions; some have been more negative, revealing that if the connexive principles of inference are to be retained, some revision of common mathematical practice must inevitably follow.

As an example of the former, Storrs McCall's [11] considers the thesis of classical set theory that the empty set is a subset of its complement and compares this with the intuition that a proposition should not entail its own negation. Indeed, [11] opens with a dialogue in which a mathematics student's plausible resistance to classical set theoretic theses serves to both illustrate and motivate connexive theses. At the conclusion of the paper, McCall demonstrates that a coherent theory of classes in which no class is contained in its complement—*i.e.*, a connexive class theory—is entirely practicable.

On the other hand, J.E. Wiredu's [26] considers what restrictions to classical set theory would have to be made in order to accommodate connexive theses. Wiredu shows that assuming connexive principles entails that even the *restricted* comprehension axioms of ZF are too strong. The upshot is that any reasonable connexive theory of sets must restrict comprehension even more severely than, say, Zermelo's

---

[1]We will adhere to Meyer's convention of, *e.g.*, [12], according to which the theory of Peano arithmetic formulated in a logic L will be labeled $L^{\sharp}$.

axiom of separation. Now, while this may demand a revision of some mathematical principles, to suggest a restriction of intuitive mathematical theses isn't uniquely offensive; following Zermelo, mathematicians—set theorists, computability theorists, *etc.*—have demonstrated a willingness to restrict comprehension if need be. In other words, while Wiredu shows the need to *restrict* some set theoretical intuitions, this does not on its face *preclude* a connexive mathematics.

Now, connexive logics share peculiar features that undoubtedly complicate matters. The *contraclassicality* of connexive logics, for example, entails that the development of connexive mathematics will be more complex—and, arguably, more interesting—than intuitionistic or substructural accounts. For example, although formally undecidable sentences in classical Peano arithmetic remain independent of its intuitionistic and relevant counterparts, there exist undecidable sentences of classical arithmetic that will become decidable modulo any reasonable connexive arithmetic.

In, *e.g.*, Peano arithmetic, there exists an undecidable sentence ξ. In the classical case, that is, when the conditional $\rightarrow$ is construed as the material conditional, ξ is equivalent to the formula $\sim(\xi \rightarrow \sim\xi)$, *i.e.*, $\sim(\sim\xi \vee \sim\xi)$. This entails that in classical Peano arithmetic the sentence $\sim(\xi \rightarrow \sim\xi)$ is likewise undecidable. Of course, in a connexive logic L and connexive arithmetic $L^{\sharp}$, one expects that $L^{\sharp}$ should prove $\sim(\xi \rightarrow \sim\xi)$ by default, witnessing that some classically undecidable statements in number theory become decidable connexively.

Although this example is extremely simple, it demonstrates that there are many subtle questions that uniquely arise in a connexive mathematics. In this paper, I wish to make a few comments on how mathematics—in particular, arithmetic—must behave if formulated connexively. We will first consider some relevant historical and philosophical topics before taking a foray into the formalization of modest subsystems of arithmetic in predicate calculi corresponding to Richard Angell's PA1 and PA2 (described in [1] and [3], respectively) and Graham Priest's P_N (described in [17]).

The more philosophical observations of this paper are guardedly optimistic. For example, we will consider a kinship between Everett Nelson's connexive theory of self-cotenability in [15] and David Nelson's philosophy of mathematics in [13] and suggest that Heinrich Wansing's work on constructive connexive logic in [21] serves as a successful harmonization of the two themes. We will also diagnose problems with Jan Łukasiewicz' argument of [9] that number theory is inconsistent with connexive principles, countering the most prominent foil to a connexive mathematics in print. In contrast, the formal results of the paper are relatively discouraging. We will see that theories of arithmetic in Angell's logics inevitably suffer from some rather counterintuitive features and that any theory including even weak induction schema (*i.e.*, those including induction for quantifier-free formulae) will have *no consequences* in Priest's system.

2 Historical Points

In this section, we will consider a few more philosophical points before proceeding to formalizations of fragments of arithmetic.

2.1 Anti-Zenonian Mathematics and *Reductiones ad Absurdum*

The suggestion of a positive philosophical foundation for connexive mathematics can be discovered in an interesting remark in William and Martha Kneales' [7]. While discussing Aristotle's articulation of connexive principles, the Kneales suggest that it is "tempting" to infer that "what [Aristotle] attacks is in effect the positive counterpart of Zeno's *reductio ad impossibile*." [7, 97] William Kneale is even more explicit in tethering Aristotle's Thesis to the rejection of *reductiones*, noting that

> the entailment assertions which Aristotle refuses to admit are just those required for justification of the hypothetical premisses in... the constructive counterpart of the *reductio ad absurdum*. [6, 66]

The connection between Aristotle's Thesis and a rejection of *reductiones* is also observed by Angell in [1], writing that "[t]he objection has been raised that such theorems... would eliminate *reductio ad absurdum* proofs." [1, 337] The importance of *reductiones* to the development of mathematics suggests that there is a mathematical thesis to be squeezed out of the connexive position.

While the Kneales are tentative about this suggestion, a tension between connexive principles and the technique of *reductio* is more also apparent in Everett Nelson's [15]. A tension with—if not outright denial of—the legitimacy of *reductiones* is implicit in his connexive analysis of entailment in [15]. Nelson notes that there exists a

> difference between a logic in which relations [of entailment] are based on facts extrinsic to the essence of the propositions and one in which they are based on the essence itself.[15, 451]

Nelson suggests that any acceptable characterization of logical entailment ought to take the latter route by analyzing the connections between the distinct essences of propositions.

One of the notable assumptions distinguishing Nelson's account of entailment can be summarized by the passage:

> If p entails q, then p is consistent with q. This assertion, together with $p \, E \, p$ [$i.e.$, the principle that p entails p], gives rise to $p \circ p$ [$i.e.$, that p is consistent with p].[15, 447]

Nelson's account of the relationship between consistency and entailment diverges from the familiar Lewisian notion of cotenability. C.I. Lewis chooses to define entailment in terms of consistency in [8],[2] so that $\varphi \to \psi$ is defined as $\sim(\varphi \circ \sim\psi)$, that is, φ entails ψ when φ and $\sim\psi$ are not mutually consistent. Hence, while Nelson infers from the thesis that $\varphi \to \varphi$ is that all propositions are self-cotenable, the Lewisian account only infers from $\varphi \to \varphi$ the much tamer claim that $\sim(\varphi \circ \sim\varphi)$, $i.e.$, that φ is inconsistent with its negation.

The identification of the *mutual consistency* of two propositions with their *cotenability* brings out some of the *prima facie* plausibility of Nelson's position. Under this reading, that every proposition is self-consistent is just to say that every proposition is tenable in some sense, which suggests that no proposition is so defective so as to preclude even its consideration. There is something admittedly attractive about this notion insofar as it comports with philosophical practice. Philosophers, for example, frequently engage in *counterpossible* reasoning, in which even inconsistent propositions can be maintained for the sake of argument. In this sense, every proposition is to some degree *tenable*—and one might reasonably identify self-cotenability with tenability *simpliciter*.

The tension between Nelson's assertion that every proposition is self-consistent and the foundation for the legitimacy of *reductiones* is subtle. If one considers a contradiction $\varphi \wedge \sim\varphi$, from this one can trivially derive a contradiction: $\varphi \wedge \sim\varphi$ itself. But if one assumes that what has been derived is *consistent* with what has been assumed, to apply a *reductio* to disprove $\varphi \wedge \sim\varphi$ is to reject $\varphi \wedge \sim\varphi$ on the basis that it entails itself.

In other words, that every proposition is self-consistent entails that within the essence of any proposition, one cannot discover any feature sufficiently defective to warrant its rejection solely on these grounds. Hence, if *reductiones* serve to reject propositions on the basis of the defectiveness of their essences, the technique of *reductio* is, in a sense, empty and vacuous.

It is interesting to note that David Nelson's [13] is also motivated by a resistance to *reductiones ad absurdum*. In the context of constructive mathematics, Nelson notes while *positive* formulae must always be constructibly verified in intuitionistic logic, there is no analogous requirement—indeed, no corresponding operation—for *constructive disproof*. The intuitionistic negation $\varphi \to \perp$ can be recognized as the assertion of the existence of a *reductio* disproof for φ. But while $\varphi \to \perp$ is a

---

[2]Lewis goes so far as to refer to the Survey System as the "Calculus of Consistencies" in [8].

perfectly well-formed formula that imparts some information about φ, the formula is *not* identified with $\sim\varphi$.

> [J]ust as in the case of an existential proposition, we may, in the case of the negation of a generality statement $\sim\forall x A(x)$, distinguish two methods of proof. In one there is presented an effective method of constructing an n such that $\sim A(n)$ is true, in the other there is presented a demonstration that $\forall x A(x)$ implies an absurdity.[13, 16–17]

If $\forall x A(x)$ is judged as false, we tend to think that there is some (possibly more than one) *false instance* $A(n)$ that is responsible for the falsity of $\forall x A(x)$.[3] Intuitionistically, to assert the truth of $\exists x A(x)$ is to have a construction of a natural number n witnessing that a verifying instance $A(n)$ is true. From a constructive standpoint, it seems just as natural to expect an effective construction of a *falsifying* instance of $A(n)$ when $\forall x A(x)$ is false. But the existence of a proof that $\forall x A(x)$ entails an absurdity does not guarantee a procedure that produces such a witness.

Hence, Nelson reasons that intuitionistic negation does not provide an adequate characterization of falsity. From this, he argues that a true commitment to constructivity in mathematics entails concern for constructive *falsity* as well as for constructive truth. The moral is that just as the intuitionist rejects non-constructive *proof* as vacuous, the intuitionist ought to reject non-constructive *disproof*, i.e., *reductiones ad absurdum*, as empty and vacuous.

The proximity between the opinions concerning *reductiones*—the connexivist rejection on the one hand and the constructivist rejection on the other—suggests a ground for a philosophy of connexive mathematics. On both accounts, a disproof of φ requires more than the mere demonstration of a defect with respect to φ. To disprove φ demands something further, such as the explicit production of evidence that φ fails to comport with arithmetical facts. On the one hand, to derive the negation of φ from φ itself is fruitless because rather than revealing a defect with respect to φ, one has deduced improperly. On the other hand, it is fruitless because such a derivation fails to reveal what makes φ false. An interesting fact is that there already exist formalisms that apparently harmonize these two considerations.

It is arguable that Wansing's approach to connexive logic described in, *e.g.*, [21] and [22] reflects this coincidence between connexivity and strong constructivity. Wansing's C and related systems are presented within a framework generalizing the semantics of David Nelson's logic of constructible falsity and, indeed, the discussion

---

[3]As an example, it is natural to assert that "all prime numbers are odd" is false *because* "2 is a prime number and is odd" is false.

of falsification in [23] is consistent with the above considerations on *reductiones ad absurdum*.

A particular feature of Wansing's C that might be thought to exhibit the convergence of the two notions is the theoremhood of formulae $\sim(A \to \bot)$ is a theorem of C, *i.e.*, the theoremhood of $\sim\neg A$ where $\neg$ denotes intuitionistic negation. On the one hand, when this feature is read as the thesis that all intuitionistically negated formulae are false in C, this might be thought to be counterintuitive. But providing a natural Brouwer-Heyting-Kolmogorov-type reading, the theoremhood of $\sim(A \to \bot)$ is interpreted as the statement that for an arbitrary formula A, it is false that there exists a procedure to convert a proof of A into a proof of $\bot$,[4] an interpretation that might be thought to capture the shared ground between Everett and David Nelson.

2.2 Łukasiewicz' Counterexample to Aristotle's Thesis

We have encountered one apparently negative result concerning connexive mathematics in Wiredu's remarks of [26] that connexive principles might require further restriction to the comprehension axioms of set theory. A more dangerous specter for connexive mathematics appears in a remark of Jan Łukasiewicz in [9], in which he aims to demonstrate the inconsistency of Aristotle's Thesis with respect to a fundamental number-theoretic principle:

> [Euclid] states first that 'If the product of two integers, a and b, is divisible by a prime number n, then if a is not divisible by n, b should be divisible by n.' Let us now suppose that $a = b$ and the product $a \times a$ (a^2) is divisible by n. It results from this supposition that 'If a is not divisible by n, then a is divisible by n.' Here we have an example of a true implication the antecedent of which is the negation of the consequent.[9, 50–51]

Let us make this argument somewhat more perspicuous by formalizing the sequence of reasoning. Where "$a \mid b$" symbolizes the relation that b is divisible by a, we provide the following scheme with n prime:

1.	$\forall a,b[n \mid (a \times b) \to [n \nmid a \to n \mid b]]$	Number-Theoretic Truth
2.	$n \mid n^2$	Number-Theoretic Truth
3.	$n \mid n^2 \to [n \nmid n \to n \mid n]$	Universal Instantiation, 1.
4.	$n \nmid n \to n \mid n$	Modus Ponens, 2,3.

---

[4]But *n.b.* that C is inherently negation inconsistent, a point that problematizes a BHK-style interpretation. Because both $\bot \to \bot$ and $\sim(\bot \to \bot)$ are theorems, for example, our naive BHK interpretation entails both the existence and non-existence of a procedure turning proofs of $\bot$ into proofs of $\bot$.

Hence, it appears that, granted very weak logical assumption, instances of Euclid's lemma serve as counterexamples to Aristotle's Thesis. Furthermore, it is not clear that Euclid's lemma can be restricted as naturally or as readily as naive comprehension.[5]

Despite Euclid's employing a conditional in the statement of the lemma, to state that if a prime n divides a composite number $a \times b$ then *either* n divides a *or* n divides b is an equally good (and perhaps superior) formulation of Euclid's lemma. If so, Łukasiewicz' argument requires an enthymematic assumption of the validity of disjunctive syllogism. Formally, the initial steps would be:

1.	$\forall a, b[n \mid (a \times b) \rightarrow [n \mid a \vee n \mid b]]$	Number-Theoretic Truth
2.	$n \mid n^2$	Number-Theoretic Truth
3.	$n \mid n^2 \rightarrow [n \mid n \vee n \mid n]$	Universal Instantiation, 1.
4.	$n \mid n \vee n \mid n$	Modus Ponens, 2,3.

If we state Euclid's lemma in this way, the most natural way to generate the counterexample to Aristotle's Thesis is by inferring $n \nmid n \rightarrow n \mid n$ from $n \mid n \vee n \mid n$, *e.g.*, by producing a conditional proof making an explicit appeal to the validity of disjunctive syllogism. But this sort of argument is much less compelling. Łukasiewicz intends to argue against Aristotle's Thesis on purely mathematical grounds, that is, Aristotle's Thesis is to be rejected not due to its inconsistency with competing logical principles, but due to its inconsistency with mathematical intuitions. Moreover, if one follows Richard Sylvan in maintaining that connexivism "coincides with the broad requirement of relevance"[18, 393], then it is virtually obligatory that one rejects disjunctive syllogism as an archetypal fallacy of relevance. Hence, the force of Łukasiewicz' counterexample to Aristotle's Thesis is not generated by features of number theory, but by assumptions concerning logic.[6]

---

[5]A referee has noted that such restrictions to Euclid's lemma—by, *e.g.*, stipulating that a and b must be either nonequal or relatively prime—would in fact stave off counterexamples to Aristotle's Thesis. It is worth mentioning that this reply shares an affinity with the approach to connexive logic described in Graham Priest's [17], in which counterexamples to Aristotle's Thesis are avoided by filtering out cases in which an antecedent is contradictory.

[6]A referee has noted that in any logic in which Weakening is admissible, $n \nmid n \rightarrow n \mid n$ will follow from the truth of $n \mid n$. This suggests that Łukasiewicz could have formulated a similar argument against Aristotle's Thesis by appeal to Weakening. Such an argument, of course, would also have to make an appeal to explicitly logical principles, something Łukasiewicz appears to be attempting to avoid in [9].

3   Fragments of Arithmetic in Connexive Logics

Łukasiewicz, I have suggested, failed to give a definitive number-theoretic refutation of Aristotle's Thesis. We have also considered some apparent connections between the connexive account of entailment and David Nelson's philosophy of mathematics. This is not to say, however, that existent connexive logics are compatible with our standard convictions concerning arithmetic. In this section, we will examine three connexive propositional logics and modest extensions thereof, revealing that any reasonable extensions of these systems are either straightforwardly incompatible with arithmetical principles or lead to severely pathological formulations of arithmetic.

3.1   Three Connexive Logics

Richard Angell's propositional logic PA1 was introduced in [1] as an attempt to capture a notion of a subjunctive conditional, in which Boethius' Thesis is presented as the *principle of subjunctive contrariety*. A further connexive logic PA2 was introduced by Angell in work first appearing as the abstract [2] and subsequently appearing in Italian as [3]. PA2 was intended to rectify certain shortcomings of PA1 and McCall's system CC1 of [10], one of which shall make an appearance in the sequel.

To define Angell's PA1 and PA2, we will follow the presentation of many-valued logics in [4].

Definition 1. *The semantic matrix for* PA1 *is* $\langle \mathscr{V}_{\mathsf{PA1}}, \mathscr{D}_{\mathsf{PA1}}, f^{\sim}_{\mathsf{PA1}}, f^{\wedge}_{\mathsf{PA1}}, f^{\rightarrow}_{\mathsf{PA1}} \rangle$ *where*

- $\mathscr{V}_{\mathsf{PA1}} = \{0, 1, 2, 3\}$
- $\mathscr{D}_{\mathsf{PA1}} = \{0, 1\}$

and the truth functions are defined by the following matrices:

$\sim$	φ
3	0
2	1
1	2
0	3

$\varphi \wedge \psi$	0	1	2	3
0	1	0	3	2
1	0	1	2	3
2	3	2	3	2
3	2	3	2	3

$\varphi \rightarrow \psi$	0	1	2	3
0	1	2	3	2
1	2	1	2	3
2	1	2	1	2
3	2	1	2	1

Definition 2. *The semantic matrix for* PA2 *is* $\langle \mathscr{V}_{\mathsf{PA2}}, \mathscr{D}_{\mathsf{PA2}}, f^{\mathsf{T}}_{\mathsf{PA2}}, f^{\sim}_{\mathsf{PA2}}, f^{\wedge}_{\mathsf{PA2}}, f^{\rightarrow}_{\mathsf{PA2}} \rangle$ *where* $\mathscr{V}_{\mathsf{PA2}} = \mathscr{V}_{\mathsf{PA1}}$ *and* $\mathscr{D}_{\mathsf{PA2}} = \mathscr{D}_{\mathsf{PA1}}$ *and the truth functions are defined by the following matrices:*

T	φ	$\sim$	φ	$\varphi \wedge \psi$	0	1	2	3	$\varphi \to \psi$	0	1	2	3
1	0	3	0	0	0	1	2	3	0	1	3	3	3
1	1	2	1	1	1	1	2	3	1	3	1	3	3
3	2	1	2	2	2	2	2	3	2	1	3	1	3
3	3	0	3	3	3	3	3	3	3	3	1	3	1

Semantic validity in both PA1 and PA2 is defined in the standard way, that is, as preservation of designated values from premises to conclusion.

Angell fails to provide any intuitive reading for these matrices. The nearest thing to a natural interpretation the matrices for PA1 might be derived from Routley and Montgomery's interpretation of McCall's CC1. With notation adjusted to reflect Angell's matrices, Routley and Montgomery write:

> CC1, for instance, can be given a semantics by associating the matrix value [0] with logical necessity, value [3] with logical impossibility, value [1] with contingent truth, and value [2] with contingent falsehood.[19, 95]

However, Routley and Montgomery concede that such an interpretation is given to *anomalies*—*e.g.*, the conjunction of two necessary truths is a contingent truth—and these anomalies follow when such an interpretation is given to PA1. It is common to treat these matrices merely as a theoretical tool. Wansing, for example, states that the semantics for CC1 "appears to be a purely formal method with little explanatory power."[21, 370]

Despite the apparent artificiality of the semantics for PA1 and PA2, Graham Priest has introduced a pair of connexive logics in [17] whose semantics are much more philosophically salient. Priest's semantics are motivated by a theme of negation as "cancellation," so that a formula $\sim\varphi$ *cancels* a formula φ. One of the formal features of this account of negation is that the cancellation of φ by $\sim\varphi$ entails that $\varphi \wedge \sim\varphi$ has no *content* and therefore entails nothing. Priest traces the provenance of this notion of negation-as-cancellation through Western philosophy, describing appearances in the work of not only Aristotle and Boethius, but also Abelard and Berkeley. In this paper, we will focus only on the "non-symmetrized" version P_N.[7]

Definition 3. *A model for* P_N *is a 3-tuple* $\mathfrak{M} = \langle W, g, v \rangle$*, where* W *is a nonempty set of points with* $g \in W$ *and* v *is a function mapping* **At** *to subsets of* W*.*

- $\mathfrak{M}, w \Vdash p$ *iff* $w \in v(p)$ *for* $p \in$ **At**

---

[7]Priest's [17] only refers to this system as the "plain connexive logic." The nomenclature P_N was introduced in [5].

- $\mathfrak{M}, w \Vdash {\sim}\varphi$ iff $\mathfrak{M}, w \nVdash \varphi$
- $\mathfrak{M}, w \Vdash \varphi \wedge \psi$ iff $\mathfrak{M}, w \Vdash \varphi$ and $\mathfrak{M}, w \Vdash \psi$
- $\mathfrak{M}, w \Vdash \varphi \vee \psi$ iff $\mathfrak{M}, w \Vdash \varphi$ or $\mathfrak{M}, w \Vdash \psi$
- $\mathfrak{M}, w \Vdash \varphi \rightarrow \psi$ iff $\begin{cases} \exists w' \in W \text{ such that } \mathfrak{M}, w' \Vdash \varphi, \text{ and} \\ \forall w' \in W, \text{ if } \mathfrak{M}, w' \Vdash \varphi \text{ then } \mathfrak{M}, w' \Vdash \psi \end{cases}$

Validity is defined by appealing to the designated world g.

Definition 4. *Validity is defined so that:*

$$\Gamma \vDash_{\mathsf{P_N}} \varphi \ \text{ if } \begin{cases} \text{there exists an } \mathfrak{M} \text{ such that for all } \psi \in \Gamma, \ \mathfrak{M}, g \Vdash \psi \\ \text{for all } \mathfrak{M} \text{ such that for all } \psi \in \Gamma, \ \mathfrak{M}, g \Vdash \psi, \text{ also } \mathfrak{M}, g \Vdash \varphi \end{cases}$$

The case of validity in $\mathsf{P_N}$ is atypical in that it is not *Tarskian*. Notably, the presumption of reflexivity (*i.e.*, self-entailment) fails, which can be clearly illustrated. Consider the set $\{p \wedge {\sim}p\}$. Then there exists no model $\mathfrak{M}$ such that $\mathfrak{M}, g \Vdash p \wedge {\sim}p$, whence $p \wedge {\sim}p \nvDash_{\mathsf{P_N}} p \wedge {\sim}p$ may be inferred.

3.2 Formal Languages

Of course, it is impossible to sufficiently express the richness of arithmetical theses in a purely propositional language. For the purposes of this section—showing that PA1, PA2, and $\mathsf{P_N}$ are questionable bases for arithmetic—we need not develop the full language of arithmetic. Rather, we appeal to weak subsystems of arithmetic, simplifying matters by considering only a modest fragment of the full language of arithmetic, *i.e.*, the language with a constant $\mathbf{0}$, a binary symbol $\doteq$ representing identity, and a successor function $(\cdot)'$.

We will define very weak languages that one may expect to be included in any sufficiently expressive language of arithmetic and then proceed to describe general schema for extending the semantics of PA1, PA2, and $\mathsf{P_N}$ to a framework rich enough to accommodate these languages.

Now, let us define the formal languages with which we will work. Suppose that we have a denumerable set **Var** of variables $\{x_0, x_1, ..., y_0, ...\}$; then the set of terms $\mathbf{Tm}^\sharp$ is

$$\{\underbrace{\tau'{\dots}'}_{n \text{ times}} \mid n \in \omega \text{ and } \tau \in \mathbf{Var} \cup \{\mathbf{0}\}\}$$

The set of closed terms $\mathbf{Tm}_C^\sharp$ is the subset comprising instances of the symbol $\mathbf{0}$ followed by finitely (possibly zero) many applications of $(\cdot)'$.

Definition 5. $\mathbf{At}^\sharp$ *is defined as the set* $\{s\dot{=}t \mid s,t \in \mathbf{Tm}^\sharp\}$ *and the set of closed atoms* $\mathbf{At}^\sharp_C$ *is defined as the set* $\{s\dot{=}t \mid s,t \in \mathbf{Tm}^\sharp_C\}$.

Given the distinct logical connectives of PA1 and PA2, we must define two languages of arithmetic.

Definition 6. *If* φ *is a formula and* $s,t \in \mathbf{Tm}^\sharp$, *then* $\varphi[s ::= t]$ *is the formula generated by replacing each instance of* s *with an instance of* t.

Definition 7. $\mathscr{L}^\sharp$ *is defined*

- *If* $\varphi \in \mathbf{At}^\sharp$ *then* $\varphi \in \mathscr{L}^\sharp$
- *If* $\varphi \in \mathscr{L}^\sharp$ *then* $\sim\varphi \in \mathscr{L}^\sharp$
- *If* $\varphi, \psi \in \mathscr{L}^\sharp$ *then* $\varphi \wedge \psi \in \mathscr{L}^\sharp$
- *If* $\varphi, \psi \in \mathscr{L}^\sharp$ *then* $\varphi \to \psi \in \mathscr{L}^\sharp$
- *If* $\varphi \in \mathscr{L}^\sharp$ *then* $\forall x(\varphi[t ::= x]) \in \mathscr{L}^\sharp$ *for* $x \in \mathbf{Var}$

Definition 8. $\mathscr{L}^\sharp_\mathsf{T}$ *is defined in a similar fashion, appending the recursive clause:*

- *If* $\varphi \in \mathscr{L}^\sharp_\mathsf{T}$ *then* $\mathsf{T}\varphi \in \mathscr{L}^\sharp_\mathsf{T}$

3.3 Protoarithmetical Theories in Angell's PA1 and PA2

In this section, we will consider some of the idiosyncrasies that will meet arithmetic if its axioms are formulated in appropriately rich extensions of PA1 and PA2.

Definition 9. *A* universal-identity (UI) extension *of* PA1 *is any deductive system extending* PA1 *rich enough to ensure that:*

- *there exist valuations* $v : \mathbf{At}^\sharp_C \to \mathscr{V}_{\mathsf{PA1}}$
- *the valuations respect the truth functions of* PA1
- *for any valuation* v *there exists a recursive method of evaluating each formula* $\forall x\varphi[t ::= x]$ *so that* $v(\forall x\varphi[t ::= x]) \in \mathscr{V}_{\mathsf{PA1}}$

A UI extension of PA2 *is defined analogously.*

We will denote these systems generically by "PA1$^+$" and "PA2$^+$," defining validity in the expected way:

Definition 10. *We say that* $\Gamma \vDash_{\mathsf{PA1}^+} \varphi$ *(respectively,* $\Gamma \vDash_{\mathsf{PA2}^+} \varphi$*) when in every* $\mathsf{PA1}^+$ *model (respectively,* $\mathsf{PA2}^+$ *model) such that* $v(\psi) \in \mathscr{D}_{\mathsf{PA1}}$ *for all* $\psi \in \Gamma$, *also* $v(\varphi) \in \mathscr{D}_{\mathsf{PA1}}$ *(respectively,* $\mathscr{D}_{\mathsf{PA2}}$*).*

A further logical definition is required before turning to look at arithmetic-specific considerations. We will define a *theory* semantically as a set of sentences closed under semantic consequence:

Definition 11. *In a deductive system* L, *a theory* T *is a collection of sentences closed under semantic consequence, that is,* T *is a set such that* $\varphi \in T$ *iff* $T \vDash_{\mathsf{L}} \varphi$.

Note an important aspect of the foregoing definition. In classical logic, the standard definition of a theory is only that $T \vDash_{\mathsf{L}} \varphi$ entails that $\varphi \in T$. While the converse holds classically by the definition of consequence—guaranteeing the felicity of the above definition—the converse must be explicitly expressed in cases in which consequence is non-Tarskian.

Now, suppose we also employ a convention defining an extension of $\mathscr{L}^\sharp$ to include formulae with open variables x, y, *etc.* Then the standard definition of a bounded universal quantifier is expressible in the theories of $\mathscr{L}^\sharp$.

Definition 12. *For a natural number* n, *define a* bounded universal quantifier *so that*

$$(\forall x \leq n)\varphi(x) =_{df} \varphi(x)[x := \mathbf{0}] \wedge \varphi(x)[x := \mathbf{0}'] \wedge ... \wedge \varphi(x)[x := \underbrace{\mathbf{0}'..'}_{n \text{ times}}]$$

A very reasonable expectation concerning bounded universal quantifiers in arithmetic is that if all natural numbers less than n have a property φ, then when $m < n$, all natural numbers less than m have the property φ.

Despite this expectation, we observe the following idiosyncrasy with respect to arithmetic in any UI extension of $\mathsf{PA1}$:

Observation 1. *There exists a formula* ψ *such that for any theory* T *in a UI extension* $\mathsf{PA1}^+$ *and natural numbers* m *and* n,

$$T \vDash_{\mathsf{PA1}^+} (\forall x \leq n + 2m)\psi(x) \rightarrow (\forall x \leq n)\psi(x)$$

although

$$T \nvDash_{\mathsf{PA1}^+} (\forall x \leq n + 2m - 1)\psi(x) \rightarrow (\forall x \leq n)\psi(x)$$

Proof. Let $x \dot{\neq} y$ denote the formula $\sim(x \dot{=} y)$ and let $\psi(x)$ denote the formula $\sim((x \dot{=} x \rightarrow (x \dot{\neq} x)) \wedge (x \dot{=} x \rightarrow x \dot{\neq} x))$. Then it can easily be confirmed that for any natural number n, the value assigned to $\psi(x)[x := \ulcorner n \urcorner]$ is 0.

Then the value assigned to $(\forall x \leq n)\psi(x)$ is determined entirely by the *parity* of n, that is, for any valuation v, we have the following:

$$v((\forall x \leq n)\psi(x)) = \begin{cases} 0 & \text{if } n \text{ is odd} \\ 1 & \text{if } n \text{ is even} \end{cases}$$

Hence, the formula $(\forall x \leq n + 2m)\psi(x) \rightarrow (\forall x \leq n)\psi(x)$ will take a value of 1 in any model of T while $(\forall \leq n + 2m - 1)\psi(x) \rightarrow (\forall x \leq n)\psi(x)$ will take a value of 2 in any model of T.

This means that for all n, $T \vDash_{\mathsf{PA1}^+} (\forall x \leq n + 2m)\psi(x) \rightarrow (\forall x \leq n)\psi(x)$ although $T \nvDash_{\mathsf{PA1}^+} (\forall x \leq n + 2m - 1)\psi(x) \rightarrow (\forall x \leq n)\psi(x)$. $\square$

Note that this is common to all $\mathsf{PA1}^+$ theories, not merely those including some fragment of arithmetic. In other words, this pathology is intimately related to John Woods' diagnosis of the "defects" of McCall's connexive $\mathsf{CC1}$. Although $\mathsf{CC1}$ is distinct from $\mathsf{PA1}$, the similarity of the two entails that Woods' objection applies equally to both systems:

> The upshot would appear to be that p connexively implies only odd-numbered conjunctions of occurrences of itself, and never even-numbered ones. [27, 474]

This feature is arguably more troubling in the context of arithmetic, as the foregoing observation lifts the pathology from matters of logical form to the behavior of bounded quantification over natural numbers.

Although the revisions to conjunction central to Angell's $\mathsf{PA2}$ seem to solve these apparently counterintuitive features of bounded quantification—$\mathsf{PA2}$ is introduced precisely to repair the pathology concerning conjunction in $\mathsf{PA1}$—there remain some peculiarities facing $\mathsf{PA2}$ theories of very weak subsystems of arithmetic. Let us define conditions for $\mathsf{PA2}^+$ theories to be *protoarithmetical*.

Definition 13. *Call a theory T* protoarithmetical *if*

- *For all $t \in \mathbf{Tm}_C^\sharp$, $t \dot{=} t \in T$*
- *For all $s, t \in \mathbf{Tm}_C^\sharp$, $s' \dot{=} t' \rightarrow s \dot{=} t \in T$*

Essentially, that T is protoarithmetical is to say that it provides a natural interpretation of identity and that each instance of the successor axiom holds.

Lemma 1. *Let v be a model of a protoarithmetical theory in a UI extension of* PA2. *Then*

- *for all $t \in \mathbf{Tm}_C^\sharp$, $v(t \doteq t) = 0$ or*
- *for all $t \in \mathbf{Tm}_C^\sharp$, $v(t \doteq t) = 1$.*

Proof. Suppose not and that there exist $s, t \in \mathbf{Tm}_C^\sharp$ such that $v(s \doteq s) = 0$ although $v(t \doteq t) = 1$. For a term $t \in \mathbf{Tm}_C^\sharp$, let $\kappa(t)$ denote the number of primes occurring in t so that, *e.g.*, $\kappa(\mathbf{0}''') = 3$.

Suppose without loss of generality that $\kappa(s) < \kappa(t)$. Then there exists a term $u \in \mathbf{Tm}_C^\sharp$ (possibly s) such that $\kappa(s) \leq \kappa(u) < \kappa(u') \leq \kappa(t)$ where $v(u \doteq u) = 0$ and $v(u' \doteq u') = 1$. Simple calculation entails that $v(u' \doteq u' \to u \doteq u) = 3$, which is not designated. This contradicts the assumed protoarithmeticity of the theory of v. $\square$

First, consider the following definitions. Let PA2$^+$ be a UI extension of PA2:

Definition 14. *A protoarithmetical* PA2$^+$ *theory T is* literal *if for every term $t \in \mathbf{At}_C^\sharp$, $T \vDash_{\mathsf{PA2}^+} (t \doteq t) \leftrightarrow \mathsf{T}(t \doteq t)$.*

Definition 15. *A protoarithmetical* PA2$^+$ *theory T is* illiterate *if for every term $t \in \mathbf{At}_C^\sharp$, $T \vDash_{\mathsf{PA2}^+} {\sim}((t \doteq t) \leftrightarrow \mathsf{T}(t \doteq t))$.*

Given the interpretation given to T in [3], a theory is literal if all statements of self-identity are literally true and a theory is illiterate if self-identity is always meant hypothetically, that is, formulae of the form $t \doteq t$ are considered absent any assumption that $t \doteq t$ is *true*.

If we use "completeness" of a theory in the model-theoretic sense, *i.e.*, that T is complete when it is the theory of some model, these are the *only* types of complete protoarithmetical PA2$^+$ theory—and *a fortiori*, the only two types of complete theories of PA2$^+$ arithmetic. Consider the following observation:

Observation 2. *For each UI extension of* PA2, *every complete protoarithmetical theory is either literal or illiterate.*

Proof. By model-theoretic completeness, T is the theory of a model v; from Lemma 1, either for all formulae $t \doteq t \in \mathbf{At}_C^\sharp$, $v(t \doteq t) = 0$ or for all $t \doteq t \in \mathbf{At}_C^\sharp$, $v(t \doteq t) = 1$. By examining the truth tables for PA2—which PA2$^+$ must respect—we draw the inference that v will either uniformly assign formulae $(t \doteq t) \leftrightarrow \mathsf{T}(t \doteq t)$ the value of 3 (in the former case) or uniformly assign such formulae the value 1. As T is the theory of v, it follows that either T is literal or illiterate. $\square$

3.4 Numerically Inductive Theories in Priest's $\mathsf{P_N}$

Just as we considered a very general scheme for extending PA1 and PA2, we can give a similar definition to suitable extensions of $\mathsf{P_N}$.

Definition 16. *A universal extension of* $\mathsf{P_N}$ *is any extension enriching* $\mathsf{P_N}$ *sufficiently so that:*

- *models have valuations* $v : \mathbf{At}_C^\sharp \to \wp(W)$

- *the forcing conditions for the connectives are identical to those in* $\mathsf{P_N}$

- *for any model there is a recursive method of evaluating a formula* $\forall x \varphi[t ::= x]$ *governing when* $w \Vdash \forall x \varphi[t ::= x]$

Recall the earlier qualification made with respect to the definition of a theory. The peculiarities of $\mathsf{P_N}$—and any first-order extensions of $\mathsf{P_N}$—have important consequences for this notion. Standardly, any set of sentences has a deductive closure modulo most deductive systems. But the notion of logical consequence in $\mathsf{P_N}$ deviates from the standard Tarskian account and not all sets of formulae can serve as the kernel of a deductive closure. For example, inasmuch as $p \wedge \sim p \nvdash_{\mathsf{P_N}} p \wedge \sim p$, the set $\{p \wedge \sim p\}$ *may not have* a deductive closure modulo $\vdash_{\mathsf{P_N}}$, *i.e.*, $\{p \wedge \sim p\}^{\vdash_{\mathsf{P_N}}} = \varnothing$. This follows intuitively from Priest's formalization of negation-as-cancellation. The set $\{p \wedge \sim p\}$ *has no content* and therefore *has no deductive closure*.

We are still concerned with showing problems with formalizing even weak arithmetics in connexive logic but will now consider theories besides protoarithmetical theories. It is likely that we would wish to retain *some* aspect of induction in our theories of arithmetic. Let us define the following notion:

Definition 17. *Let T be a theory in a UI extension of* $\mathsf{P_N}$ *whose language extends* $\mathscr{L}^\sharp$. *Then T is* numerically inductive *if the signature of T extends the signature of arithmetic and for every formula $\varphi(x)$, the formula*

$$[(\varphi(\mathbf{0}) \wedge \forall x(\varphi(x) \to \varphi(x'))) \to \forall x \varphi(x)] \in T.$$

Observation 3. *In any UI extension of* $\mathsf{P_N}$ *there are no numerically inductive theories.*

Proof. Let ψ be an arbitrary formula of the language (*e.g.*, $\mathbf{0} \doteq \mathbf{0}$), let $\varphi(x) = (\psi \to \sim\psi)$, and let $\mathsf{P_N^+}$ be an arbitrary UI extension of $\mathsf{P_N}$. Then for no w in any model $\mathfrak{M}$ will $\mathfrak{M}, w \Vdash \varphi(\mathbf{0})$. Thus for arbitrary formulae ξ and ζ, the formula $(\psi(\mathbf{0}) \wedge \xi) \to \zeta$

will be logically false. Hence, no matter how quantification is handled in $\mathsf{P_N^+}$—*i.e.*, irrespective of the points at which $\forall x(\varphi(x) \to \varphi(x'))$ and $\forall x(\varphi(x))$ are true—$\varphi(\mathbf{0}) \wedge \forall x(\varphi(x) \to \varphi(x'))$ will hold at no w in the model by the forcing conditions for connexive implication. It follows the formula $(\varphi(\mathbf{0}) \wedge \forall x(\varphi(x) \to \varphi(x'))) \to \forall x \varphi(x)$ will hold at no w and *a fortiori* at no designated point g. Moreover, if $(\varphi(\mathbf{0}) \wedge \forall x(\varphi(x) \to \varphi(x'))) \to \forall x \varphi(x) \in T$ then the deductive closure of T is $\varnothing$, that is, there are no theories including all instances of the arithmetical induction schema. $\qquad\qquad\qquad\qquad\qquad\qquad\qquad\qquad\qquad\qquad\qquad\qquad\qquad\square$

We noted that the scheme $(\varphi(\mathbf{0}) \wedge \xi) \to \zeta$ will have logically false instances for all ξ and ζ in any UI extension of $\mathsf{P_N}$. Thus, even significant restrictions to induction—*e.g.*, induction on quantifier-free formulae—cannot obtain. However, it is worth noting that this pathology does not at first blush conflict with Robinson's Q, as Q lacks any type of induction axiom scheme. Hence, advocates of very weak subsystems of Peano arithmetic may not be discouraged by this observation.

Furthermore, it is worth mentioning that this constitutes a trivial decidability result concerning the Peano axioms in $\mathsf{P_N^+}$, *i.e.*, $(\mathsf{P_N^+})^\sharp$. Consider an arbitrary formula φ in a language extending $\mathscr{L}^\sharp$ and ask: Is φ a logical consequence of the theory $(\mathsf{P_N^+})^\sharp$? Of course, we have an answer: No. By Observation 3, $(\mathsf{P_N^+})^\sharp$ has *no* logical consequences and is thus trivially decidable.

On its face, Observation 3 might seem to entail that first-order extensions of $\mathsf{P_N}$ are not suitable bases for the formulation of arithmetic. Despite this, it is important to note that the fact that *Peano arithmetic* cannot be formulated in this system does not entail that *arithmetic* cannot be so formulated. All this shows is that systems of arithmetic that include even meager species of induction are not practicable. But not all theories of arithmetic posit induction. Observation 3 fails to rule out that Robinson's Q—the axioms of Peano arithmetic without the induction schema—can be formulated in Priest's system. Moreover, several philosophical standpoints *anticipate* that inclusion of the induction axioms with the other Peano axioms *should* yield a trivial result. From the perspective of strict finitism—a position with which Priest himself has flirted in [16]—the true pathology is found not in the *failure* of induction but in the supposition that it *holds*. That an arithmetic with induction is *empty* in the sense of Observation 3, after all, is a side of the same medal as *e.g.*, Edward Nelson's attempts to show that arithmetic with induction is *inconsistent* (see, *e.g.*, [14] for Edward Nelson's criticism of the inherent impredicativity of the induction schema).

4   Conclusion

In this paper, we have surveyed a number of topics concerning the implementation of a connexive arithmetic. The foregoing observations have ranged from the encouraging (*e.g.*, the rebuttal to Łukasiewicz' argument against Aristotle's Thesis) to the discouraging (*e.g.*, the pathologies of PA1). However, especially in the cases of UI extensions of PA2 and P_N, what has been uncovered is that conjoining connexive and arithmetical concerns yields a landscape that is not insuperable, although its terrain may appear quite alien from the perspective of classical mathematics.

One can justifiably interpret this as an invitation to further study. Supposing that there exist models of, say, Q in UI extensions of PA2 and P_N, there are many natural questions that emerge: Is there a robust and natural way of interpreting the distinction between literal and illiterate models in $PA2^+$? Is there any recursive restriction of numerical induction (just as we restrict comprehension) that would permit induction in P_N^+ arithmetic by, *e.g.*, considering only induction for negation-free formulae?

Given the pathologies that greet arithmetic formulated in these systems, it is fair to say that the supposition that there exist a model of even Q in these systems is a much stronger assumption than the existence of a model for classical PA. In this regard, Wansing's C provides a bit of a ray of light with respect to the pursuit of connexive arithmetic. Given the faithful embedding of quantified C into positive intuitionistic logic described in [21], it seems very likely that $C^\sharp$ is Post consistent relative to Heyting arithmetic $J^\sharp$, that is:

Conjecture 1. *If $J^\sharp$ has a model then there exists a model of $C^\sharp$.*

This conjecture is very plausible and suggests that in the case of C, the presumption of the existence of a model of $C^\sharp$ will be a corollary of mathematical orthodoxy.

As a field, modern studies in connexive logic have generally struggled with reconciling the *prima facie* plausibility of connexive theses with the pathologies that emerge during their formalization. Connexive arithmetic, it appears, faces a derivative problem. Nevertheless, the connexive milieu contains the possibility of a very distinct philosophy of mathematics and, more specifically, may yield a novel way of looking at arithmetic. Pursuing these matters further may be interesting not only for the sake of connexive arithmetic itself, but as a means of providing a new perspective on classical mathematical practice.

References

[1] R. B. Angell. A propositional logic with subjunctive conditionals. *Journal of Symbolic Logic*, 27(3):327–343, 1962.

[2] R. B. Angell. Three logics of subjunctive conditionals (Abstract). *Journal of Symbolic Logic*, 32(4):556–557, 1967.

[3] R. B. Angell. Tre logiche dei condizionali congiuntivi. In C. Pizzi, editor, *Leggi di Natura, Modalità, Ipotesi*, pages 156–180. Feltrinelli, Milan, 1978.

[4] L. Bolc and P. Borowik. *Many-Valued Logics*. Springer, New York, 1992.

[5] T. M. Ferguson. Logics of nonsense and Parry systems. *Journal of Philosophical Logic*, 44(1):65–80, 2015.

[6] W. Kneale. Aristotle and the consequentia mirabilis. *Journal of Hellenic Studies*, 77(1):62–66, 1957.

[7] W. Kneale and M. Kneale. *The Development of Logic*. Oxford University Press, Oxford, 1962.

[8] C. I. Lewis. *A Survey of Symbolic Logic*. University of California Press, Berkeley, CA, 1918.

[9] J. Łukasiewicz. *Aristotle's Syllogistic from the Standpoint of Modern Formal Logic*. Oxford University Press, Oxford, first edition, 1951.

[10] S. McCall. Connexive implication. *Journal of Symbolic Logic*, 31(3):415–433, 1966.

[11] S. McCall. Connexive class logic. *Journal of Symbolic Logic*, 32(1):83–90, 1967.

[12] R. K. Meyer. *Arithmetic Formulated Relevantly*. Canberra, 1976. Unpublished manuscript.

[13] D. Nelson. Constructible falsity. *Journal of Symbolic Logic*, 14(1):16–26, 1949.

[14] E. Nelson. *Predicative Arithmetic*. Princeton University Press, Princeton, 1986.

[15] E. J. Nelson. Intensional relations. *Mind*, 39(156):440–453, 1930.

[16] G. Priest. Is arithmetic consistent? *Mind*, 103(411):337–349, 1994.

[17] G. Priest. Negation as cancellation and connexive logic. *Topoi*, 18(2):141–148, 1999.

[18] R. Routley. Semantics for connexive logics, I. *Studia Logica*, 37(4):393–412, 1978.

[19] R. Routley and H. Montgomery. On systems containing Aristotle's Thesis. *Journal of Symbolic Logic*, 33(1):82–96, 1968.

[20] J. K. Slaney, R. K. Meyer, and G. Restall. Linear arithmetic desecsed. *Logique et Analyse*, 39(155–156):379–387, 1996.

[21] H. Wansing. Connexive modal logic. In R. Schmidt, I. Pratt-Hartmann, M. Reynolds, and H. Wansing, editors, *Advances in Modal Logic*, volume 5, pages 367–383. Kings College Publications, London, 2005.

[22] H. Wansing. Constructive negation, implication, and co-implication. *Journal of Applied Non-Classical Logics*, 18(2–3):341–364, 2008.

[23] H. Wansing. Proofs, disproofs, and their duals. In L. Beklemishev, V. Goranko, and V. Shehtman, editors, *Advances in Modal Logic*, volume 8, pages 483–505. Kings College

Publications, London, 2010.

[24] Z. Weber. Transfinite numbers in paraconsistent set theory. *Review of Symbolic Logic*, 3(1):71–92, 2010.

[25] Z. Weber. Transfinite cardinals in paraconsistent set theory. *Review of Symbolic Logic*, 5(2):269–293, 2012.

[26] J. E. Wiredu. A remark on a certain consequence of connexive logic for Zermelo's set theory. *Studia Logica*, 33(2):127–130, 1974.

[27] J. Woods. Two objections to system CC1 of connexive implication. *Dialogue*, 7(3):473–475, 1968.

Received October 2015

Beyond System **P** – Hilbert-Style Convergence Results for Conditional Logics with a Connexive Twist

Matthias Unterhuber
University of Pittsburgh, United States
mau29@pitt.edu

Abstract

The paper has three aims. Firstly, the convergence result of conditional logics for Systems **P** and **R** is extended; based on a Hilbert style axiomatization it is proved that Lewis's System **V** and Burgess's variant system, System **V***, are nothing but Lehmann and Magidor's System **R**. Secondly, it is shown that connexive principles are the center stage of axiomatizations of System **P** and System **R**. They introduce a proof-theoretic dependency of two core principles of System **P** and System **R** – Cautious Monotonicity and Rational Monotonicity – even when the connexive principles are formulated as default rules. Thirdly, the impossibility result for an extension of classical conditional logics by unrestricted connexive principles is strengthened. It is shown that such an impossibility result ensues even when Principle Refl is given up, where the latter asserts that 'if A then A' is a theorem. As a consequence, on pain of inconsistency any complete classical conditional logic can include connexive principles only in a restricted form, where classical conditional logics are minimal conditional logics that take classical propositional calculus to govern propositional connectives other than conditionals. Implications of the strengthened impossibility result are discussed.

In the last decades there has been a strong convergence result for logics describing conditionals. Again and again, two proof-theoretic systems, System **P** and System **R**, have been obtained by utilizing a number of different semantic concepts. System

I would like thank Heinrich Wansing, Thomas Ferguson, Hitoshi Omori, Hannes Leitgeb, Graham Priest, Claudio Pizzi, Igor Douven, and Hans Rott as well as two anonymous referees. Special thanks goes to Kevin Kelly for our discussions from which I profited greatly. I acknowledge financial support by the Center for Mathematical Philosophy at the University of Munich and the Center for Philosophy of Science at the University of Pittsburgh. The visiting fellowship at the University of Munich was made possible by the Humboldt Foundation on behalf of Hannes Leitgeb.

R is the stronger of the two systems and differs from System **P** by including the following principle, where 'RM' is short for 'Rational Monotonicity' and $\neg$, $\wedge$, $\Box\!\!\rightarrow$ abbreviate 'negation', 'conjunction', and the conditional operator, respectively:

RM   if $p\,\Box\!\!\rightarrow r$ and $\neg(p\,\Box\!\!\rightarrow \neg q)$ then $(p \wedge q)\,\Box\!\!\rightarrow r$

Both systems, System **P** and System **R**, have been found to characterize the conservative core of material consequence relations of default logics and non-monotonic logics [36, 14, 15] – consequence relations that describe warranted inferences given any type of assumption, including default assumptions. Inferences based on default assumptions do not guarantee that the inferences hold under all possible circumstances; rather they might have to be retracted in light of further evidence.

Both systems also govern probabilistic inferences when the consequence relation is understood in terms of preservation of high probability, independently of the particular framework chosen ([1, 2, 3, 4, 32, 10]; [25, Ch. 10]; [16, Theorems 69–71 and 97–100]). In addition, Systems **P** and **R** have been found to be the conditional logics that result from reconstructing full beliefs – an agent does or does not believe a given proposition – based on a probabilistic degree of belief framework in a non-trivial way ([18, 19, 22]; see also [13]). The principles of System **P** also receive empirical support as principles guiding reasoning with conditionals. Studies in human reasoning show that humans employ the rules of System **P** rather than those of classical logic in probabilistic reasoning tasks (e.g., [27]).

Observe that not all of the convergence results put System **P** and System **R** in opposition. In an Adams type probabilistic framework (e.g., [1, 2, 3, 4, 25, 32, 16]) both systems result from the same probabilistic semantics.[1] Whether System **P** or System **R** is obtained is determined by the degree to which the language is restricted in which conditionals are expressed. System **P** results when the object language neither allows for nestings of conditionals (e.g., no 'if p then if q then r') nor for Boolean combinations of conditionals (e.g., no 'if p then q, and if r then s'). In Adams type semantics arbitrary nestings of conditionals and arbitrary Boolean combinations of conditionals cannot be allowed, for otherwise Lewis's [21] triviality result ensues. Yet, we can extend the language without triviality by including either negated conditionals or disjunctions with conditionals as disjuncts and in both cases System **R** is obtained ([4, 32, 16]; cf. [39, Ch. 3.6]). However, we cannot include both, negated conditionals and disjunctions with conditionals as disjuncts, for otherwise Lewis's triviality result kicks in ([21]; cf. [39, Ch. 3.7]).[2]

---

[1]Note that not all probabilistic conditional logic systems validate the rules of System **P**. Systems **CLD** ('CLD' for 'Conditional Logic System of Douven'; [7, Theorem 5.2.1]) and **O** (e.g., [12, 34]) do not.

[2]Lewis's triviality result is due to the assumption that the probability of a conditional 'if p then

Notice that not all convergence results are of this sort. For example, Lin and Kelly's [22] account of full beliefs differs from Leitgeb's [18, 19] and this difference gives rise to System **P** rather than System **R** (see also [13]).

The present paper pursues three goals. Firstly, it extends the convergence result by proof theoretic means and shows that Lewis's [20] counterfactual conditional logic **V** and Burgess's [5] variant system, System **V***, are nothing but System **R** of Lehmann and Magidor's [15]. The only requirement for the equivalence result of Systems **R**, **V**, and **V*** is that the language is expressive enough to either allow negated conditionals or disjunctions with conditionals as disjuncts (otherwise some of the axioms cannot be expressed in the language). In fact, the result still upholds if the language is extended to a full conditional language – in line with modal conditional logics such as Lewis's – which admit arbitrary nestings and Boolean combinations of conditionals.

The equivalence of Systems **R** and **V** is shown in Theorem 7.32 of [39], whereas the proof the equivalence of **R** and **V*** is new. Restating the equivalence result in this paper seems worthwhile, since this result has only been stated in a 300+ pages book and has not yet received attention in the literature. In addition, the equivalence proofs differ. This paper employs a full Hilbert style axiomatization of the respective logics, devoid of axiom schemata. This contrasts with [39] who uses a mixture of natural deduction and Hilbert-style axiomatization. A full Hilbert style axiomatization is used here for the following reasons: Firstly, the resulting axiomatization seems to be more natural than alternative axiomatizations (e.g., [35, 39]). Secondly, it is a further aim of this paper to show that a full Hilbert style axiomatization of conditional logics is feasible as a means for the investigation of proof theoretic dependencies in such systems. Thirdly, the use of the current Hilbert style axiomatization makes the reliance of the equivalence proofs on the tautologies as described by classical propositional calculus more explicit.

Moreover, the new result – the equivalence of Burgess's System **V*** and System **R** – is particularly interesting. For the axiomatization of System **V*** Burgess uses an alternative version of RM. This version of RM uses conditionals of the form (1) $\alpha \vee \beta \mathbin{\square\!\!\rightarrow} \neg\beta$ rather than conditionals of the form (2) $\alpha \vee \beta \mathbin{\square\!\!\rightarrow} \alpha$. In conditional logics, such as Lewis's, conditionals of type (1) are stronger than conditionals of type (2). (2) indicates not only that α is as normal as β – as does (1) – but asserts that α is strictly more normal than β (cf. [31, p. 1224]). The equivalence proof also

---

q' is equal to the conditional probability $P(q \mid p)$, which is satisfied in Adams type semantics. Note that Lewis's triviality result does not generalize to non-probabilistic semantics for conditionals, such as Lewis's [20] (cf. [39, Ch. 3.7]). In addition, there are probabilistic semantics that differ from Adams's approach and which allow for conjunctions of conditionals without falling prey to Lewis's triviality result either (e.g., [11]).

makes explicit the conditions under which a specification of RM in terms of such a strict normality ordering is viable.

Note that System **P** comes in two different brands. One brand ([3, 22]) includes the default version of Boethius's thesis – a connexive principle – which can be formulated as follows ('DBT' for 'Default Version of Boethius's Thesis'):[3] [4]

DBT   if α is consistent then $(\alpha \,\square\!\!\rightarrow \beta) \rightarrow \neg(\alpha \,\square\!\!\rightarrow \neg\beta)$

DBT merits the label 'default' for the following reason. By default we seem to restrict BT intuitively to cases in which the antecedent α is consistent.

In contrast to the first brand of System **P**, the Principle DBT is missing from the second brand of System **P**, as described by [14] and [1, 2] (see [39, Ch. 3.6]).[5] This default formulation of BT contrasts with the full version of BT and a full version of Aristotle's Thesis (AT), which are as follows ('$\rightarrow$' for 'material implication', [42, 43, 23]):

AT $\neg(p \,\square\!\!\rightarrow \neg p)$
BT $(p \,\square\!\!\rightarrow q) \rightarrow \neg(p \,\square\!\!\rightarrow \neg q)$

Analogously to DBT, also a default version of DAT such as the following can be introduced:

DAT   if α is consistent then $\neg(\alpha \,\square\!\!\rightarrow \neg\alpha)$

It might appear inconsequential whether a connexive principle, such as DAT or DBT, is included in the axiomatization of Systems **P** and **R**. However, it is not, and it is the second aim of the paper to show why this is not so. Observe that there is an alternative way to restrict AT and BT. This approach uses the notion of possibility rather than the notion of consistency and we shall discuss this second approach at the end of this section.

On the contrary, connexive principles are highly relevant for the axiomatization of System **P** and System **R**, since they connect two core principles of Systems **P** and **R**, Principle RM and Principle CM ('Cautious Monotonicity'), where CM is a theorem of both, System **P** and System **R**:

---

[3]Adams's [3, p. 56] explicitly states that $\alpha \,\square\!\!\rightarrow \beta$ and $\alpha \,\square\!\!\rightarrow \beta$ are inconsistent in his logic. Adams's assertion might look like a full version of BT. However, it is a proper default version of BT as [3, p. 45f] allows only conditionals with consistent antecedents to be expressible in the language (cf. [39, Section 3.6.3]). In contrast, Adams's [3] official rule is in fact closer to DAT (see below) than to DBT. [3, p. 61, R7] states that if α is logically consistent and $\alpha \wedge \beta$ is logically inconsistent then $\alpha \,\square\!\!\rightarrow \beta$ implies anything in his logic.

[4]For an empirical investigation of AT and BT in a probabilistic context see, for example, [26].

[5]There is a further difference between axiomatizations of **P**; some axiomatizations of System **P** include the centering axioms [1, 2, 3, 22] whereas others, in particular [14], do not. I restrict myself here to the discussion of System **P** as described in [14] plus possibly Principles DBT and DAT.

RM $\quad (p\,\square\!\!\!\rightarrow r) \rightarrow (\neg(p\,\square\!\!\!\rightarrow \neg q) \rightarrow ((p \wedge q)\,\square\!\!\!\rightarrow r))$

CM $\quad (p\,\square\!\!\!\rightarrow r) \rightarrow ((p\,\square\!\!\!\rightarrow q) \rightarrow ((p \wedge q)\,\square\!\!\!\rightarrow r))$

In particular, it can be shown that given very weak assumptions CM is derivable from RM in the presence of DAT [DBT], whereas the remaining axioms of System **P** and RM do not imply CM when DAT [DBT] is absent. The only 'substantial' assumption for this result is that the conditional logic **L** renders the following principle a theorem ('Refl' for 'Reflexivity'):

Refl $\quad p\,\square\!\!\!\rightarrow p$

Consequently, connexive principles are at the center stage for any axiomatization of System **R**. This result is heightened in the light of the equivalence of Lewis's **V** and Burgess's System **V*** and **R** – implying that connexive principles play a decisive role for a broad range of conditional logics.

Before turning to the third aim of the paper, let me explain why DAT and DBT are of interest in a probabilistic semantics for conditionals as described above. The core idea of such probabilistic semantics for conditionals is that conditionals are acceptable iff their conditional probability is high. In contrast to classical and modal logics the validity of arguments is not understood as truth preservation – the truth of all premises guarantees that the conclusion is true as well – but rather preservation of high (conditional) probability. Preservation of high (conditional) probability can be spelled out in very different ways. For example, in an Adams type semantics the premises of a valid argument warrant an arbitrarily high probability of the conclusion. Despite the different ways in which preservation of high (conditional) probability can be understood, the same proof-theoretic systems result as outlined above.

A peculiarity in such a high conditional probability framework is that AT and BT are almost valid. Ignoring cases in which conditional probabilities are trivial in a sense described below, $P(\beta\,|\,\alpha)$ and $P(\neg\beta\,|\,\alpha)$ cannot both be high for a given probability function P. This gives rise to the fact that we cannot accept both $\alpha \,\square\!\!\!\rightarrow \beta$ and $\alpha \,\square\!\!\!\rightarrow \neg\beta$, which is exactly what BT asserts.[6] The only exception occurs when the respective conditional probability $P(\cdot\,|\,\alpha)$ is trivial in the sense that $P(\gamma\,|\,\alpha) = 1$ for any γ. On a standard account of conditional probabilities this amounts to the unconditional probability of α equalling zero.[7] For example, in [3] the validity of

---

[6] Analogously, $P(\alpha\,|\,\alpha)$ and $P(\neg\alpha\,|\,\alpha)$ cannot both be high and since $P(\alpha\,|\,\alpha) = 1$ it follows that $P(\neg\alpha\,|\,\alpha)$ is low.

[7] Standard accounts of conditional probability define conditional probability in terms of unconditional probability. The above notion of trivial conditional probability functions still applies on a Popper function account of conditional probability (e.g., [41]). On such an account conditional probability is taken as primitive.

DAT and DBT is achieved by treating conditional probabilities $P(\beta \mid \alpha)$ as undefined whenever $P(\alpha)$ equals zero. (Adams employs a standard account of conditional probability.) However, in order for Adams's semantics to work, the conditional probability $P(\beta \mid \alpha)$ has to be defined for some probability function P. The only case in which this is not possible is the case in which α is a logical contradiction. This is due to the fact that the standard axioms of probability require any logical contradiction to receive a probability value of zero (as they require any logical truth to receive a probability value of one). The default versions of AT and BT – DAT and DBT – now exclude this problematic case, due to requirement that the antecedent of a conditional has to be consistent in order for AT and BT to be applicable.

The third and final aim of this paper is to strengthen the impossibility result for connexive extensions of classical conditional logics. The impossibility result asserts that any classical conditional logic can be extended only by restricted versions of AT and BT, on pain of inconsistency of the system. Classical conditional logics are minimal conditional logics, where minimal conditional logics require (i) consequents of conditionals to form deductively closed sets and (ii) conditionals with a fixed consequent and logically equivalent antecedents are logically equivalent. Classical conditional logics are then minimal conditional logics which take classical logic to govern logical equivalence and deductive closure as described in (i) and (ii) for propositional connectives other than conditionals.

The original impossibility result is stated for any classical conditional logic that includes Refl. Let us see why this impossibility result holds. Take $\bot$ ('falsum') to be any contradiction, for example, $p \wedge \neg p$. Then, by Refl $\bot \,\Box\!\!\rightarrow\, \bot$ follows. However, $\bot \,\Box\!\!\rightarrow\, \bot$ implies by (i) above that $\bot \,\Box\!\!\rightarrow\, \neg\bot$ for classical conditional logics, thus contradicting both BT and AT. On the other hand, $\bot$ is the only case in which this inference holds up. Thus, if we ensure that the antecedent is consistent – as done by DBT and DAT – then the resulting system will not be inconsistent.

The paper strengthens this impossibility result by proving that this result holds without Refl. To this end, Chellas-Segerberg (CS) semantics [6, 35] is used, a semantics which is sound and complete with respect to the class of classical conditional logic. It is shown that AT and BT correspond to distinct frame conditions in that semantics. Remarkably, AT corresponds to a strictly stronger frame condition than BT, yet the frame condition corresponding to AT cannot be satisfied in CS semantics. As a result, even in such a minimal classical conditional logic as described by CS semantics no extension by connexive principles without qualification is possible. Note that the great majority of conditional logics discussed in literature are classical. This includes Systems $\mathbf{P}$ and $\mathbf{R}$ described above.[8]

---

[8]Non-classical conditional logic systems with a probabilistic semantics are, for example, System

The 'culprit' of the strengthened impossibility result is in fact the principle LT ('logical truth') which asserts that for any theorem β, $\alpha \mathbin{\square\!\!\rightarrow} \beta$ is a theorem. The original and the strengthened triviality results suggest then two possible approaches. We can restrict AT and BT as for example [3] and [22] do – or we can restrict Refl and LT, where [20, Sect. 1.6] and [29] seem to take the latter approach. (More on this in Section 5).

To describe the proof-theoretic relation between CM and RM due to DAT [DBT], the proof theory is extended to include default rule inferences. Default rules are based on consistency conditions apart from derivability conditions. Note that consistency conditions are nothing but non-provability conditions in disguise.[9] To see that, take the consistency condition in DAT and DBT. (1) By definition α is consistent in a classical conditional logic **L** iff α is not inconsistent in **L**. (2) A formula α is inconsistent in **L** iff there exists a proof in **L** that shows that $\neg\alpha$ is a theorem of **L**. (1) and (2) imply that a formula α is consistent in a logic **L** iff $\neg\alpha$ is not provable in **L**, yielding the desired result.[10] To make the structure of proofs more perspicuous, I shall give the proofs in terms of non-derivability conditions rather than consistency conditions.

In general, default logics – logics which admit default rules – are non-monotonic. That is, whenever a formula α is derivable from a set of formulae Δ, it is not guaranteed that α is derivable from a set Γ such that $\Delta \subseteq \Gamma$. Monotonicity may fail due to reliance on default rules, since a derivation of α from Δ by a default rule depends on the non-derivability of a formula β from the formulae in Δ – a condition which may not be satisfied for Γ. However, there is a set of theorems of default logics which relies on non-trivial default rules, yet is monotonic in the above

---

CLD ([7]; see my Footnote 1) and System **O** (e.g., [12, 34]). The main difference between these logics and classical conditional logics is that these non-classical conditional logics do not satisfy AND (see Section 1). Interestingly, Douven's logic neither satisfies Refl.

[9]The notion of consistency employed here is nothing but the standard notion of consistency from logic (e.g., [8, p. 119]; see Section 1.2).

[10]The use of non-derivability conditions makes the resulting system a proper refutation logic (cf. [37]). Note that the inclusion of default rules leads to problems. For a general first-order extension of such systems, either (i) finite axiomatizability has to be given up (by including an infinite number of non-theorems as axioms) or (ii) both non-derivability and derivability have to axiomatized. Step (ii) is also problematic as it implies that there is no guarantee that such a first-order extension is complete (e.g., [33, pp. 42–44]). Note, however, that some first-order extensions are complete, such as negation-complete first-order theories [39, p. 53f]. Observe furthermore that the axiomatization of DAT and DBT depends only on the non-provability of single formulae rather than the non-derivability of formulae from sets of formulae [39, p. 50f] – a feature that may make all the difference for a completeness result. Both notions diverge since the non-derivability of a formula from a set of formulae is *not* reducible to the non-provability of a single formula – in contrast to provability and derivability.

383

sense. This happens to be the case if a formula α can be derived based on both – a non-provability condition for some formula β – and a provability claim of the same formula β. In order to allow for this type of 'stable' inference, a new inference rule, TNT ('TNT' for 'Theorem Non-Theorem'), is introduced that governs this type of inference. Such an extensions of a default logic proof theory is required for this paper, as the proof-theoretic dependence between CM and RM based on DAT [DBT] is of that sort.

Moreover, to establish the negative claim – that in the absence of DAT [DBT] the proof-theoretic dependence of CM and RM does not uphold – CS semantics is used. To this end, the proof draws on the correspondence result of the conditional logic principles of Systems **P** and **R** in CS semantics [39, 40].

Before we turn to the outline of the paper let us discuss an alternative way to restrict AT and BT. This approach requires the antecedent α in AT and BT to be possible (short: $\Diamond \alpha$) rather than to be consistent, where possibility is the stronger of the two notions. This gives rise to the following two restricted versions of AT and BT:

PAT $\quad \Diamond \alpha \to \neg(\alpha \,\Box\!\!\!\to\, \neg\alpha)$
PBT $\quad \Diamond \alpha \to ((\alpha \,\Box\!\!\!\to\, \beta) \to \neg(\alpha \,\Box\!\!\!\to\, \neg\beta))$

However, PAT and PBT are too weak to capture AT and BT in a meaningful way. In conditional logics $\Diamond \alpha$ is customarily defined in such a way that $\Diamond \alpha$ iff $\neg(\alpha \,\Box\!\!\!\to\, \neg\alpha)$.[11] [12] [13] Thus, $\Diamond \alpha$ is tantamount to AT, trivializing PAT and PBT. (PBT is trivial when a classical conditional logic is used, as AT implies BT in classical conditional logics by Lemma 5 of Section 2.) The problem with this definition of $\Diamond \alpha$ is that it is too strong. However, weaker definitions are hard to come by if not impossible. In fact, if we understand $\Diamond \alpha$ asserting that any formula α is true at some world in some intensional model, we end up with the requirement that α is consistent, as described by DAT [DBT]. The same holds for probabilistic models, such as Adams's, when 'possibly α' is understood as it being possible that α is assigned an arbitrarily

---

[11] In conditional logics $\Box \alpha$ is defined as $\neg\alpha \,\Box\!\!\!\to\, \alpha$ (e.g., [20, p. 22]; [24, p. 52]). Furthermore, $\Diamond \alpha$ is the dual of $\Box \alpha$ implying $\Diamond \alpha$ iff $\neg\Box\neg\alpha$. By condition (ii) of classical conditional logics, the above equality follows. Note that in order to warrant the interpretation of $\Diamond \alpha$ and $\Box \alpha$ as 'possibly α' and 'necessarily α' in the usual sense, respectively, the classical conditional logic has to be augmented by Principles Refl and MOD, where MOD is as follows: $(\neg p \,\Box\!\!\!\to\, p) \to (q \,\Box\!\!\!\to\, p)$.

[12] [20, p. 22] employs a second type of possibility, called 'inner possibility' (short: $\Diamond\!\!\!\cdot\, \alpha$), where $\Diamond\!\!\!\cdot\, \alpha$ can be defined as $(\top \,\Box\!\!\!\to\, \bot) \vee \neg(\top \,\Box\!\!\!\to\, \neg\alpha)$. However, using $\Diamond\!\!\!\cdot\, \alpha$ rather than $\Diamond \alpha$ for PAT and PBT does not work. PAT and PBT are designed to exclude the cases in which α is logically false, whereas $\Diamond\!\!\!\cdot\, \alpha$ holds trivially for logically false formulae α.

[13] [38] also uses the notion of possibility in his conditional logic semantics, as described by his accessibility relation R. It is, however, easy to see that in his semantics R is formally ineffective and thus can be omitted [20, p. 78].

high probability. As a result, we shall restrict ourselves to versions DAT and DBT rather than PAT and PBT, respectively.

The paper proceeds as follows. Section 1 describes the proof theory and model theory used throughout the paper. Section 2 strengthens the impossibility result of connexive principles in classical conditional logics. Section 3 proves the equivalence of Systems **R**, **V**, and **V*** and Section 4 characterizes the proof-theoretic relation between CM and RM due to connexive principles such as DBT. Finally, Section 5 discusses some implications of the present results.

1 Basic Definitions

This section describes the proof theory and model theory used in this paper. To this end, two languages are introduced – the object language of the conditional logic and the language in which the frame conditions of the model theory are formulated. Furthermore, the traditional proof theory used in Sections 2 and 3 are described and the model theoretic notions employed by CS semantics are outlined. The extension of this proof theory to the default logic used for the derivation of CM from RM is postponed to Section 4.

1.1 Languages

Throughout this paper I shall employ the full conditional logic language $\mathfrak{L}_{\text{CL}}$ ('CL' for 'Conditional Logic'). $\mathfrak{L}_{\text{CL}}$ contains atomic propositional variables $p, q, r, s, t, p_1,$ $p_2, \ldots \in \text{AV}$ (AV is the set of atomic propositional variables) and is closed under the truth-functional propositional connectives $\neg$ ('negation'), $\rightarrow$ ('material implication'), $\wedge$ ('conjunction'), $\vee$ ('disjunction'), and $\leftrightarrow$ ('material coimplication') as well as the two-place modal operator $\Box\!\!\rightarrow$ ('conditional'). In particular, language $\mathfrak{L}_{\text{CL}}$ allows for nestings and Boolean combinations of conditional formulae.

The expressions $\alpha, \alpha_1, \alpha_2, \ldots, \beta, \gamma, \ldots$ stand for to arbitrary formulae of language $\mathfrak{L}_{\text{CL}}$. Henceforth, I shall refer by 'formulae' to formulae of language $\mathfrak{L}_{\text{CL}}$, except when indicated otherwise. I also omit outer parentheses and further parentheses as I assume that $\neg$ binds stronger than $\wedge$ and $\vee$ and that $\wedge$ and $\vee$ bind stronger than $\Box\!\!\rightarrow$ and $\rightarrow$. For example, $p \wedge q \Box\!\!\rightarrow r$ is short for $((p \wedge q) \Box\!\!\rightarrow r)$.

In addition to $\mathfrak{L}_{\text{CL}}$ I employ the language $\mathfrak{L}_{\textbf{FC}}$ ('FC' for 'Frame Condition') – the language in which the structural conditions for our semantics are formulated. Language $\mathfrak{L}_{\textbf{FC}}$ is the fragment of the first order logic language which contains (a) variables for worlds $w, w', w'', \ldots$ and sets of worlds $X, Y, X_1, X_2, \ldots$, (b) connectives $\sim$ ('negation'), $\curlywedge$ ('conjunction'), $\curlyvee$ ('disjunction'), and $\Rightarrow$ ('material implication'), (c) quantifiers $\forall$ ('universal quantifier'), and $\exists$ ('existential quantifier'), and (d) the non-logical predicate $R(w, w', X)$ ('accessibility relation'), where w, w' are arbitrary world in a given set of possible worlds W and X is a subset of W. Furthermore, I shall abbreviate '$R(w, w', X)$' by '$wR_X w'$'.

1.2 Proof Theory

Propositional Logic   The axiomatization of the propositional part of the conditional logic is given by the following axioms and rules ('A' for 'axiom', 'R' for 'rule'):

A1	$p \rightarrow (q \rightarrow p)$	
A2	$(p \rightarrow (q \rightarrow r)) \rightarrow ((p \rightarrow q) \rightarrow (p \rightarrow r))$	
A3	$(\neg p \rightarrow \neg q) \rightarrow (q \rightarrow p)$	
R1	If α and $\alpha \rightarrow \beta$ then β	MP
R2	From α to conclude $\alpha[p_1/\beta_1, \ldots, p_n/\beta_n]$	Sub
R3	if α then $\alpha[\beta_1//\gamma_1, \ldots, \beta_n//\gamma_n]$, such that $\beta_1, \ldots, \beta_n$ and $\gamma_1, \ldots, \gamma_n$ are substitution instances of the definienda and definiens of the following definitions, respectively, or vice versa:	

$$p \wedge q =_{\mathrm{df}} \neg(p \rightarrow \neg q)$$
$$p \vee q =_{\mathrm{df}} \neg p \rightarrow q$$
$$p \leftrightarrow q =_{\mathrm{df}} \neg((p \rightarrow q) \rightarrow \neg(q \rightarrow p))$$

The axiomatization is a variant of the well known Frege-Hilbert axiomatization of classical propositional logic. 'MP' and 'Sub' abbreviate 'Modus Ponens' and 'Substitution', respectively. The formula $\alpha[p_1/\beta_1, \ldots, p_n/\beta_n]$ is the formula which results from simultaneously substituting every instance of $p_1, \ldots, p_n$ in α by an instance of $\beta_1, \ldots, \beta_n$, respectively. In contrast, $\alpha[\beta_1//\gamma_1, \ldots, \beta_n//\gamma_n]$, is any formula α which results from replacing some formula β_i, $1 \leq i \leq n$, by γ_i for some occurrence of β_i in α. I refer to rules that are only admissible by the phrase 'from ... to conclude ...'. Proper rules are described by 'if ... then ...'. In semantic terms, admissible rules only guarantee that if the premises are valid, so is the consequent. In contrast, proper rules are also truth preserving. Whenever the premises are true, so is the conclusion. For example, due to MP we can infer that if p and $p \rightarrow q$ are true, so is q. In contrast, by Sub we can only infer that if α is valid, so are its substitution instances. This difference is also mirrored by the fact that MP renders $p \rightarrow ((p \rightarrow q) \rightarrow q)$ a theorem, whereas there is no such theorem for Rule Sub.

A set of formulae $\mathbf{L}$ is a logic (in the traditional sense) if it contains A1–A3 and is closed under rules R1–R3. A proof of a formula α in such a logic $\mathbf{L}$ is a finite sequence of formulae ending with α such that every formula in that sequence is an axiom of $\mathbf{L}$ or follows by the rules of $\mathbf{L}$ from preceding formulae. A formula

α is provable in **L** (short: $\vdash_{\mathbf{L}} \alpha$) iff there exists a proof of α in **L**. A formula α is derivable from Γ in **L** (short: $\Gamma \vdash_{\mathbf{L}} \alpha$) iff $\vdash_{\mathbf{L}} \beta_1 \wedge \ldots \wedge \beta_n \to \alpha$ for some n such that $\beta_1, \ldots, \beta_n \in \Gamma$. Furthermore, a formula α is **L**-consistent iff $\{\alpha\} \nvdash_{\mathbf{L}} \bot$, where $\nvdash_{\mathbf{L}}$ denotes non-derivability in logic **L** and $\bot$ is defined as $p \wedge \neg p$. In the present proof system rules require other formulae to be provable. To show that such rules are derivable we thus have to presuppose that (other) formulae are provable. To indicate that the derivability of a formula depends on provability of such a formula, I shall introduce such a provability assumptions and mark them with the phrase 'Condition of ...'. Steps in a proof that depend on such a provability assumption are marked by a plus sign and the number of the line on which the derivation depends. An example of such a derivation is the proof of Lemma 14. Dependence of provability assumptions contrast with non-derivability assumptions which are marked by a minus sign. Proofs and derivations based on non-provability assumptions are described in more detail in Section 4.

A set of formulae Γ is **L**-inconsistent if $\Gamma \vdash_{\mathbf{L}} \bot$, where $\bot$ can be defined as $p \wedge \neg p$. A formula α is **L**-inconsistent iff $\{\alpha\} \vdash_{\mathbf{L}} \bot$. A logic **L** is inconsistent iff for every set of formulae Γ it follows that $\Gamma \vdash_{\mathbf{L}} \bot$. A set of formulae [a formula, a logic] is **L**-consistent iff it is not **L**-inconsistent [**L**-inconsistent, inconsistent]. This standard notion of consistency ensures that a formula α is **L** inconsistent iff $\neg \alpha$ is a theorem of **L**. This implies that α is **L**-consistent iff $\neg \alpha$ is not a theorem of **L** [8, p. 119] and it is this latter equivalence which I shall frequently employ. Furthermore, I shall omit reference to **L** when no ambiguity arises.

In addition to this extension of Hilbert style proofs, I simplify proofs in the following two ways: Firstly, any theorem of propositional calculus can be added as a new line in a proof with the label 'pc' (propositional calculus). Secondly, I use the following derived rules:

DR1   if $p \to q$ and $q \to r$ then $p \to r$
DR2   if $p \to (q \to r)$ and $r \to s$ then $p \to (q \to s)$
DR3   if $p \to q$ and $r \to (q \to s)$ then $r \to (p \to s)$
DR4   if $p \to q$ and $p \to (q \to r)$ then $p \to r$
DR5   if $p \to q$ and $r \to (q \to s)$ then $p \to (r \to s)$
DR6   if $p \to (\neg q \to r)$ and $p \to (q \to s)$ then $p \to (\neg s \to r)$

Classical Conditional Logic   Let **L** be a logic. **L** is then a classical conditional logic iff it contains the following axiom and is closed under the following rules:

A4	$(p \Box\!\!\rightarrow q) \rightarrow ((p \Box\!\!\rightarrow r) \rightarrow (p \Box\!\!\rightarrow q \land r))$	AND
R4	from $\beta \rightarrow \gamma$ to conclude $(\alpha \Box\!\!\rightarrow \beta) \rightarrow (\alpha \Box\!\!\rightarrow \gamma)$	RW
R5	from $\alpha \rightarrow \beta$ and $\beta \rightarrow \alpha$ to conclude $(\alpha \Box\!\!\rightarrow \gamma) \rightarrow (\beta \Box\!\!\rightarrow \gamma)$	LLE
R6	from β to conclude $\alpha \Box\!\!\rightarrow \beta$	LT

'RW', 'LLE', and 'LT' abbreviate 'Right Weakening', 'Left Logical Equivalence' and 'Logical Truth', respectively. The weakest classical conditional conditional is also known as System **CK**. The present axiomatization of System **CK** differs from alternative axiomatizations [6, 35, 39]. Firstly, it is a Hilbert axiomatization that avoids axiom schema and uses axioms rather than rules whenever possible – in contrast to [6] and [35]. In particular, Rule LT is used rather than the following axiom schema which is employed by [35] (see also [39]), where $\top$ might be defined as $p \lor \neg p$:

LT* $\alpha \Box\!\!\rightarrow \top$

LT is more natural than LT* for the following reason: LT* suggests that there is something special about the constant $\top$ that allows it to be inferred as a consequent β of a conditional $\alpha \Box\!\!\rightarrow \beta$ with an arbitrary antecedent α. This, however, is not the case. Rather, $\top$ is a placeholder that allows us to infer conditionals with arbitrary antecedents if the consequent is a theorem of the respective logic. This is, however, exactly what LT asserts.

    Secondly, the present axiomatization replaces AND, RW, and LT by the following rule employed by [6]:

RCK   from $\beta_1 \land \ldots \land \beta_n \rightarrow \gamma$ to conclude $(\alpha \Box\!\!\rightarrow \beta_1) \land \ldots \land (\alpha \Box\!\!\rightarrow \beta_n) \rightarrow (\alpha \Box\!\!\rightarrow \gamma)$, $n \geq 0$

Note that in the case $n = 0$, RCK reduces to LT. That both axiomatizations are equivalent can be seen from Lemma 13.

    Using AND, RW, and LT rather than RCK has the following advantage: Not only do the object language proofs become simpler. AND, RW, and LT assert distinct properties of conditionals. For example, RW and LT require (other) formulae to be theorems, whereas AND does not. Using them as separate principles rather than the composite rule RCK allows us to make the role of those principles in the equivalence proofs more explicit.

    Let us now return to the default formulation of the connexive principles. DAT and DBT are the default rule formulations of AT and BT, respectively, and are described as follows:

DAT   If $\not\vdash \neg\alpha$ then $\neg(\alpha \Box\!\!\rightarrow \neg\alpha)$
DBT   If $\not\vdash \neg\alpha$ then $(\alpha \Box\!\!\rightarrow \beta) \rightarrow \neg(\alpha \Box\!\!\rightarrow \neg\beta)$

I specify rules with non-provability assumptions in such a way that only non-provability assumptions are explicitly marked by '$\nvdash$' – in contrast to regular provability conditions and conclusions.

1.3 Model Theory

As model theory serves CS semantics,[14] I shall restrict myself to the bare minimum of semantic notions needed to obtain the desired results, where $Pow(X)$ is the power set of a set X:

Definition 1. $\langle W, R \rangle$ is a Chellas frame iff
(a) W is a non-empty set of possible worlds
(b) $R \subseteq W \times W \times Pow(W)$

Definition 2. Let $\langle W, R \rangle$ be a Chellas frame. $\langle W, R, V \rangle$ is then a Chellas model iff V is a valuation function such that $V \colon \mathcal{L} \to Pow(W)$ and the following conditions hold for all $w \in W$:
(a) $\langle W, R, V \rangle \models_w \neg\alpha$ iff $\langle W, R, V \rangle \not\models_w \alpha$
(b) $\langle W, R, V \rangle \models_w \alpha \to \beta$ iff if $\langle W, R, V \rangle \models_w \alpha$ then $\langle W, R, V \rangle \models_w \beta$
(c) $\langle W, R, V \rangle \models_w \alpha \Box\!\to \beta$ iff for all $w' \in W$ such that $w R_{V(\alpha)} w'$ it holds that
$$\langle W, R, V \rangle \models_{w'} \beta$$

The expressions $\langle W, R, V \rangle \models_w \alpha$ and $\langle W, R, V \rangle \not\models_w \alpha$ stand for $w \in V(\alpha)$ and $w \notin V(\alpha)$ for a given model $\langle W, R, V \rangle$, respectively.

A formula α is a valid in a Chellas model $\langle W, R, V \rangle$ (short: $\langle W, R, V \rangle \models \alpha$) iff for all its substitution instances β and all worlds $w \in W$ it is the case that $\langle W, R, V \rangle \models_w \beta$. A formula α is a valid on a Chellas frame $\langle W, R \rangle$ (short: $\langle W, R \rangle \models \alpha$) iff for all Chellas models $\langle W, R, V \rangle$ it holds that $\langle W, R, V \rangle \models \alpha$. A formula α follows from a set of formulae Γ in a class of Chellas model $\mathbb{M}$ (short: $\Gamma \models_{\mathbb{M}} \alpha$) iff for all worlds w in all Chellas models $\langle W, R, V \rangle$ in $\mathbb{M}$ the following holds: if $\langle W, R, V \rangle \models_w \beta$ for all $\beta \in \Gamma$ then $\langle W, R, V \rangle \models_w \alpha$. A set of formulae Γ is satisfiable in a Chellas model $\langle W, R, V \rangle$ iff there is a world w in W such that $\langle W, R, V \rangle \models_w \alpha$ for all $\alpha \in \Gamma$. A set of formulae Γ is satisfiable in a class of Chellas models $\mathbb{M}$ iff Γ is satisfiable in a Chellas model $\langle W, R, V \rangle$ in $\mathbb{M}$. A classical conditional logic **L** is complete with respect to a class of Chellas models $\mathbb{M}$ iff for all sets of formulae Γ and all formulae α it holds that if $\Gamma \models_{\mathbb{M}} \alpha$ then $\Gamma \vdash_{\mathbf{L}} \alpha$. A formula α corresponds to frame condition C_α iff for all Chellas frames $\langle W, R \rangle$ it holds: $\langle W, R \rangle \models \alpha$ iff $\langle W, R \rangle$ satisfies C_α.

---

[14]A more detailed account of CS semantics can be found in [39] (see also [40]), including a general, non-trivial completeness proof for a lattice of systems described by thirty conditional logic principles and corresponding structural conditions.

2   Strengthening the Impossibility Result

We observed earlier that any classical conditional logic that includes Refl cannot be extended to include either AT and BT, on pain of inconsistency. Before describing the relationship between AT, BT and their variants as well as strengthening this impossibility result to systems without Refl, observe that Refl is not warranted by all interpretations of conditionals. For example, Refl should not be a theorem when $p \mathbin{\Box\!\!\rightarrow} q$ is understood as conditional obligation in the sense of 'if p is factually the case, then q ought to be the case'. A second such interpretation is given when $p \mathbin{\Box\!\!\rightarrow} q$ describes an agent believing q after his/her beliefs are revised by p. Such a revision process might not be successful, leading to the rejection of p rather than its acceptance.

It can be shown that the following variants of AT and BT, AT′ and BT′, are theorems of classical conditional logics, whenever AT and BT, respectively, are:

AT′ $\neg(\neg p \mathbin{\Box\!\!\rightarrow} p)$

BT′ $(\neg p \mathbin{\Box\!\!\rightarrow} q) \rightarrow \neg(\neg p \mathbin{\Box\!\!\rightarrow} \neg q)$

This result is described by the following lemma:

Lemma 3. AT [BT] is a theorem of a classical conditional logic whenever AT′ [BT′] is, and vice versa.

Proof. I only give the derivation of AT′ from AT. The proofs of the remaining facts are analogous.

1. $\neg(p \mathbin{\Box\!\!\rightarrow} \neg p)$	AT
2. $\neg(\neg p \mathbin{\Box\!\!\rightarrow} \neg\neg p)$	1, Sub $(p/\neg p)$
3. $p \rightarrow \neg\neg p$	pc
4. $(\neg p \mathbin{\Box\!\!\rightarrow} p) \rightarrow (\neg p \mathbin{\Box\!\!\rightarrow} \neg\neg p)$	3, RW
5. $(p \rightarrow q) \rightarrow (\neg q \rightarrow \neg p)$	pc
6. $((\neg p \mathbin{\Box\!\!\rightarrow} p) \rightarrow (\neg p \mathbin{\Box\!\!\rightarrow} \neg\neg p)) \rightarrow (\neg(\neg p \mathbin{\Box\!\!\rightarrow} \neg\neg p) \rightarrow \neg(\neg p \mathbin{\Box\!\!\rightarrow} p))$	5, Sub $(p/\neg p \mathbin{\Box\!\!\rightarrow} p, q/\neg p \mathbin{\Box\!\!\rightarrow} \neg\neg p)$
7. $\neg(\neg p \mathbin{\Box\!\!\rightarrow} \neg\neg p) \rightarrow \neg(\neg p \mathbin{\Box\!\!\rightarrow} p)$	4, 6, MP
8. $\neg(\neg p \mathbin{\Box\!\!\rightarrow} p)$	2, 7, MP

$\square$

Furthermore, the following lemma establishes that any version of AT implies any version of BT:

Lemma 4. BT and BT′ are theorems of a classical conditional logic whenever AT or AT′ are.

Proof. By Lemma 5 BT is a theorem of a classical conditional logic whenever AT is.[15] Lemma 3 establishes that AT is a theorem of such a logic whenever AT' is, and likewise for BT and BT'. This yields the desired result. $\square$

Lemma 5. If AT then BT.
Proof.

1. $(p\Box\!\!\to q)\to((p\Box\!\!\to r)\to(p\Box\!\!\to q\wedge r))$ AND
2. $(p\Box\!\!\to\neg q)\to((p\Box\!\!\to q)\to(p\Box\!\!\to\neg q\wedge q))$ 1, Sub $(q/\neg q, r/q)$
3. $\neg q\wedge q\to\neg p$ pc
4. $(p\Box\!\!\to\neg q\wedge q)\to(p\Box\!\!\to\neg p)$ 3, RW
5. $(p\Box\!\!\to\neg q)\to((p\Box\!\!\to q)\to(p\Box\!\!\to\neg p))$ 2, 4, DR2
6. $(p\to(q\to r))\to(\neg r\to(q\to\neg p))$ pc
7. $((p\Box\!\!\to\neg q)\to((p\Box\!\!\to q)\to(p\Box\!\!\to\neg p)))\to$ 6, Sub $(p/p\Box\!\!\to\neg q, q/p\Box\!\!\to q, r/p\Box\!\!\to\neg p)$

   $(\neg(p\Box\!\!\to\neg p)\to((p\Box\!\!\to q)\to\neg(p\Box\!\!\to\neg q)))$
8. $\neg(p\Box\!\!\to\neg p)\to((p\Box\!\!\to q)\to\neg(p\Box\!\!\to\neg q))$ 5, 7, MP

$\square$

In contrast, the converse of Lemma 5 cannot be established. Only the following, weaker result holds:

Lemma 6. From BT to conclude $(p\Box\!\!\to p)\to\neg(p\Box\!\!\to\neg p)$
Proof.

1. $(p\Box\!\!\to q)\to\neg(p\Box\!\!\to\neg q)$ BT
2. $(p\Box\!\!\to p)\to\neg(p\Box\!\!\to\neg p)$ 1, Sub $(q/\neg p)$

$\square$

Lemma 7. Let **L** be a classical conditional logic that renders Refl a theorem. Then, (i) AT is a theorem of **L** iff BT is and (ii) DAT is a derivable rule of **L** iff DBT is.
Proof. The lemma is a direct consequence of Lemmata 5 and 6. $\square$

To see that the converse of Lemma 5 is not valid, note that the variants of AT and BT correspond to distinct frame conditions in CS semantics.

Lemma 8. AT and BT correspond to frame condition C_{AT} and C_{BT}, respectively, where C_{AT} and C_{BT} are as follows:

C_{AT} $\forall X\forall w\exists w'(wR_Xw'\curlywedge w'\in X)$
C_{BT} $\forall X\forall w\exists w'(wR_Xw')$

Proof. See Appendix A. $\square$

---

[15]Note that Lemma 5 is stronger than required as it ensures that BT is true whenever AT is.

Observe that in spite of the correspondence result the following lemma can be given concerning C_{AT}:

Lemma 9. There is no Chellas frame $\langle W, R \rangle$ which satisfies C_{AT}.
Proof. Let $\langle W, R \rangle$ be an arbitrary Chellas frame. It holds trivially that $\varnothing \subseteq W$, yet there is no world $w' \in \varnothing$. A fortiori, there is no world $w \in W$ such that $w R_{\varnothing} w'$ and $w' \in \varnothing$. Consequently, $\langle W, R \rangle$ does not satisfy C_{AT}. □

Thus, although AT corresponds to a structural condition in CS semantics, this structural condition is never satisfied in the semantics. Observe that the failure of AT is due to the use of the empty set. The only formulae that have to be assigned the empty set by all valuation functions are the inconsistent formulae. Thus, AT is bound to fail when we substitute p by $\bot$. This observation gives rise to the following fact, where $\top$ is defined as $\neg\bot$:

Fact 10. There is no world w in a Chellas model $\langle W, R, V \rangle$ such that $\langle W, R, V \rangle \models_w \neg(\bot \mathbin{\square\!\!\rightarrow} \top)$.

Based on this fact, the following theorem can be proved:

Theorem 11. Any classical conditional logic **L** extended by AT is either inconsistent or incomplete.
Proof. Let **L** be a logic extended by AT. By a standard proof (e.g., [8, p. 135]) a classical conditional logic **L** is complete with respect to a class of Chellas models $\mathbb{M}$ iff every **L**-consistent set of formulae is satisfiable in the class of Chellas models $\mathbb{M}$. Note that $\neg(\bot \mathbin{\square\!\!\rightarrow} \top)$ is a substitution instance of AT and thus a **L** theorem of any set of formulae Γ. By a standard proof it can be shown that given Γ is **L**-consistent, so is $\Gamma \cup \{\neg(\bot \mathbin{\square\!\!\rightarrow} \top)\}$. Assume that Γ is **L**-consistent and **L** is complete with respect to the class of Chellas models. Completeness implies that $\neg(\bot \mathbin{\square\!\!\rightarrow} \top)$ is satisfiable, which it is not, due to Fact 10. Contradiction. As this holds for any set of formulae Γ it follows that **L** is either inconsistent or incomplete. □

Let me now give a diagnosis of the strengthened impossibility result. The 'culprit' is LT, as described in classical conditional logics. By LT it follows that $\bot \mathbin{\square\!\!\rightarrow} \neg\bot$, whereas AT implies $\neg(\bot \mathbin{\square\!\!\rightarrow} \neg\bot)$ yielding a contradiction. The result is surprising, since on the one hand LT is often thought to be an innocuous, almost trivial principle. On the other hand, consistent connexive logics target other principles, such as RW ([30, Footnote 14]; [43, Sect. 2.1]).

The impossibility result leaves two avenues for classical conditional logics to validate connexive principles. To obtain a consistent conditional logic, we can either restrict AT [BT] or alternatively Refl [LT]. We saw in the introductory section that

[3] and [22] take that former approach. However, it is also possible to use the following restricted versions of Refl and LT:

DRefl if α is consistent, then $\alpha \mathbin{\square\!\!\rightarrow} \alpha$

DLT if α is consistent and β is a theorem, then $\alpha \mathbin{\square\!\!\rightarrow} \beta$

As we shall see in Section 5, [29] and [20, Sect. 1.6] take that approach. However, instead of a the requirement of the antecedent α to be consistent, the authors require α to be possible, giving rise to the following principles:

PRefl $\Diamond\alpha \rightarrow (\alpha \mathbin{\square\!\!\rightarrow} \alpha)$

PLT if β is a theorem, then $\Diamond\alpha \rightarrow (\alpha \mathbin{\square\!\!\rightarrow} \beta)$

3   Equivalence of Systems R, V, and V*

As starting point of the equivalence proofs of **R**, **V**, and **V*** serves System **P** as described by [14]. This system is a classical conditional logic **L** which in addition includes the following principles ([14, Definition 5.1]; [39, Theorem 7.17]):

Refl $p \mathbin{\square\!\!\rightarrow} p$

CM $(p \mathbin{\square\!\!\rightarrow} r) \rightarrow ((p \mathbin{\square\!\!\rightarrow} q) \rightarrow (p \wedge q \mathbin{\square\!\!\rightarrow} r))$

Or $(p \mathbin{\square\!\!\rightarrow} r) \rightarrow ((q \mathbin{\square\!\!\rightarrow} r) \rightarrow (p \vee q \mathbin{\square\!\!\rightarrow} r))$

System **R** differs from System **P** by including the following additional principle [15, Definition 3.4]:

RM $(p \mathbin{\square\!\!\rightarrow} r) \rightarrow (\neg(p \mathbin{\square\!\!\rightarrow} \neg q) \rightarrow (p \wedge q \mathbin{\square\!\!\rightarrow} r))$

Note that this axiomatization of System **P** and System **R** is devoid of any connexive principles. To simplify the Hilbert style equivalence proofs I shall use the following principle which is a theorem of any classical conditional logic containing Refl:

Refl$^+$ $(p \mathbin{\square\!\!\rightarrow} q) \rightarrow (p \mathbin{\square\!\!\rightarrow} p \wedge q)$ [Refl]

In general, principles in brackets indicate which principles beyond those of classical conditional logics are needed for the respective theorem to be provable. Let us prove Refl$^+$.

Lemma 12. From Refl to conclude Refl$^+$.

Proof.

1. $(p \mathbin{\square\!\!\rightarrow} q) \rightarrow ((p \mathbin{\square\!\!\rightarrow} r) \rightarrow (p \mathbin{\square\!\!\rightarrow} q \wedge r))$ AND
2. $(p \mathbin{\square\!\!\rightarrow} p) \rightarrow ((p \mathbin{\square\!\!\rightarrow} q) \rightarrow (p \mathbin{\square\!\!\rightarrow} p \wedge q))$ 1, Sub $(q/p, r/q)$
3. $p \mathbin{\square\!\!\rightarrow} p$ Refl
4. $(p \mathbin{\square\!\!\rightarrow} q) \rightarrow (p \mathbin{\square\!\!\rightarrow} p \wedge q)$ 3, 2, MP

$\square$

3.1 Equivalence of Systems V and R

System **V** is the basic conditional logic system of Lewis. It is defined as the minimal logic **L** which includes the following axioms and is closed under the following rules ([20, p. 132f]; [17, p. 54f]):

Axioms:
Refl $p \,\square\!\!\rightarrow p$
MOD $(\neg p \,\square\!\!\rightarrow p) \rightarrow (q \,\square\!\!\rightarrow p)$
LV $(p \,\square\!\!\rightarrow \neg q) \vee ((p \wedge q \,\square\!\!\rightarrow r) \leftrightarrow (p \,\square\!\!\rightarrow (q \rightarrow r)))$

Rules:
RCK* from $\beta_1 \wedge \ldots \wedge \beta_n \rightarrow \gamma$ to conclude $(\alpha \,\square\!\!\rightarrow \beta_1) \rightarrow (\ldots \rightarrow ((\alpha \,\square\!\!\rightarrow \beta_n) \rightarrow (\alpha \,\square\!\!\rightarrow \gamma)) \ldots)$, $n \geq 1$

LE   Exchange of Logical Equivalents:
LE1 from $\alpha \rightarrow \beta$ and $\beta \rightarrow \alpha$ to conclude $(\alpha \,\square\!\!\rightarrow \gamma) \rightarrow (\beta \,\square\!\!\rightarrow \gamma)$
LE2 from $\beta \rightarrow \gamma$ and $\gamma \rightarrow \beta$ to conclude $(\alpha \,\square\!\!\rightarrow \beta) \rightarrow (\alpha \,\square\!\!\rightarrow \gamma)$

'RCK*', 'LE', and 'Mod' abbreviate 'Rule of System CK' – a rule which is characteristic of all classical conditional logics – as well as 'Exchange of Logical Equivalents', and 'Modality', respectively. Note that RCK* – unlike RCK – excludes the case in which $n = 0$. Moreover, LV is the axiom which is characteristic of Lewis's System **V**. To make the equivalence proof of System **V** and System **R** more perspicuous, I split LV into the following halves:

LV1 $(p \wedge q \,\square\!\!\rightarrow r) \rightarrow (\neg(p \,\square\!\!\rightarrow \neg q) \rightarrow (p \,\square\!\!\rightarrow (q \rightarrow r)))$
LV2 $(p \,\square\!\!\rightarrow (q \rightarrow r)) \rightarrow (\neg(p \,\square\!\!\rightarrow \neg q) \rightarrow (p \wedge q \,\square\!\!\rightarrow r))$

To further simplify the equivalence proof I shall use a divide and conquer strategy. Before giving the equivalence proof, the following will be shown first: (i) LE1, RCK*, and Refl are equivalent to LLE, RW, AND, and Refl (Lemma 13), (ii) RM and LV2 are equivalent for classical conditional logics (Lemma 18), and (iii) the following principle is equivalent to Principle Or for any classical conditional logic rendering Refl a theorem (Lemma 21):

S $(p \wedge q \,\square\!\!\rightarrow r) \rightarrow (p \,\square\!\!\rightarrow (q \rightarrow r))$

Note that LE2 is a trivial consequence of RW and is not needed for the following result:

Lemma 13. From RCK*, LE1, LE2, and Refl to conclude AND, RW, LLE, and Refl, and vice versa.
Proof. LE1 is nothing but LLE. Lemmata 14–16 give then the desired result. $\square$

Lemma 14. From RCK* to conclude RW.
Proof.

1. $\beta \rightarrow \gamma$		Condition of RW
2. $(\alpha \Box\!\!\rightarrow \beta) \rightarrow (\alpha \Box\!\!\rightarrow \gamma)$	+1	1, RCK* $(n = 1)$

$\square$

Lemma 15. From RCK* to conclude AND.
Proof.

1. $\beta \wedge \gamma \rightarrow \beta \wedge \gamma$		pc
2. $(\alpha \Box\!\!\rightarrow \beta) \rightarrow ((\alpha \Box\!\!\rightarrow \gamma) \rightarrow (\alpha \Box\!\!\rightarrow \beta \wedge \gamma))$	+1	RCK* $(n = 2)$

$\square$

Lemma 16. From AND and RW to conclude RCK*.
Proof.

1.	$\beta_1 \wedge \ldots \wedge \beta_n \rightarrow \gamma$	Condition of RCK*
2.	$(\alpha \Box\!\!\rightarrow \beta_1 \wedge \ldots \wedge \beta_n) \rightarrow (\alpha \Box\!\!\rightarrow \gamma)$ +1	1, RW
3.	$(p \Box\!\!\rightarrow q) \rightarrow ((p \Box\!\!\rightarrow r) \rightarrow (p \Box\!\!\rightarrow q \wedge r))$	AND
4.	$(\alpha \Box\!\!\rightarrow \beta_1 \wedge \ldots \wedge \beta_{n-1}) \rightarrow ((\alpha \Box\!\!\rightarrow \beta_n) \rightarrow (\alpha \Box\!\!\rightarrow \beta_1 \wedge \ldots \wedge \beta_n))$	3, Sub $(p/\alpha, q/\beta_1 \wedge \ldots \wedge \beta_{n-1}, r/\beta_n)$
5.	$(\alpha \Box\!\!\rightarrow \beta_1 \wedge \ldots \wedge \beta_{n-1}) \rightarrow ((\alpha \Box\!\!\rightarrow \beta_n) \rightarrow (\alpha \Box\!\!\rightarrow \gamma))$ +1	4, 2, DR2

$$\vdots$$

(2n+1).	$(\alpha \Box\!\!\rightarrow \beta_1) \rightarrow (\ldots \rightarrow ((\alpha \Box\!\!\rightarrow \beta_n) \rightarrow (\alpha \Box\!\!\rightarrow \gamma)) \ldots)$ +1	5, (n-2)x Sub on 3 and (n-2)x DR2

$\square$

Lemma 17. System **V** is a classical conditional logic.
Proof. Due to Lemma 13 it only remains to be shown that from RW and Refl to conclude LT as given by the following derivation:

1. β		Condition of LT
2. $\beta \rightarrow (\alpha \rightarrow \beta)$		pc
3. $\alpha \rightarrow \beta$	+1	1, 2, MP
4. $(\alpha \Box\!\!\rightarrow \alpha) \rightarrow (\alpha \Box\!\!\rightarrow \beta)$	+1	3, RW
5. $p \Box\!\!\rightarrow p$		Refl
6. $\alpha \Box\!\!\rightarrow \alpha$		5, Sub (p/α)
7. $\alpha \Box\!\!\rightarrow \beta$	+1	6, 4, MP

$\square$

Lemma 18. Let **L** be a classical conditional logic. RM is in **L** iff LV2 is.
Proof. By Lemmata 19 and 20. □

Lemma 19. From RM to conclude LV2.
Proof.

1. $(p\,\square\!\!\!\rightarrow r)\rightarrow(\neg(p\,\square\!\!\!\rightarrow\neg q)\rightarrow(p\wedge q\,\square\!\!\!\rightarrow r))$ RM
2. $(p\,\square\!\!\!\rightarrow(q\rightarrow r))\rightarrow(\neg(p\,\square\!\!\!\rightarrow\neg q)\rightarrow(p\wedge q\,\square\!\!\!\rightarrow(q\rightarrow r)))$ 1, Sub $(r/q\rightarrow r)$
3. $(p\,\square\!\!\!\rightarrow q)\rightarrow(p\,\square\!\!\!\rightarrow p\wedge q)$ Refl$^+$ [Refl]
4. $(p\wedge q\,\square\!\!\!\rightarrow(q\rightarrow r))\rightarrow(p\wedge q\,\square\!\!\!\rightarrow p\wedge q\wedge(q\rightarrow r))$ 3, Sub $(p/p\wedge q,q/q\rightarrow r)$
5. $p\wedge q\wedge(q\rightarrow r)\rightarrow r$ pc
6. $(p\wedge q\,\square\!\!\!\rightarrow p\wedge q\wedge(q\rightarrow r))\rightarrow(p\wedge q\,\square\!\!\!\rightarrow r)$ 5, RW
7. $(p\wedge q\,\square\!\!\!\rightarrow(q\rightarrow r))\rightarrow(p\wedge q\,\square\!\!\!\rightarrow r)$ 4, 6, DR1
8. $(p\,\square\!\!\!\rightarrow(q\rightarrow r))\rightarrow(\neg(p\,\square\!\!\!\rightarrow\neg q)\rightarrow(p\wedge q\,\square\!\!\!\rightarrow r))$ 2, 7, DR2

□

Lemma 20. From LV2 to conclude RM.
Proof.

1. $(p\,\square\!\!\!\rightarrow(q\rightarrow r))\rightarrow(\neg(p\,\square\!\!\!\rightarrow\neg q)\rightarrow(p\wedge q\,\square\!\!\!\rightarrow r))$ LV2
2. $r\rightarrow(q\rightarrow r)$ pc
3. $(p\,\square\!\!\!\rightarrow r)\rightarrow(p\,\square\!\!\!\rightarrow(q\rightarrow r))$ 2, RW
4. $(p\,\square\!\!\!\rightarrow r)\rightarrow(\neg(p\,\square\!\!\!\rightarrow\neg q)\rightarrow(p\wedge q\,\square\!\!\!\rightarrow r))$ 3, 1, DR1

□

Lemma 21. Let **L** be a classical conditional logic which has Refl as a theorem. Then, Or is a theorem of **L** iff S is.
Proof. By Lemmata 22 and 26. □

Lemma 22. From S to conclude Or.
Proof.

1. $(p\,\square\!\!\!\rightarrow r)\rightarrow(p\vee q\,\square\!\!\!\rightarrow(p\vee\neg q\rightarrow r))$ Lemma 23 [S]
2. $(q\,\square\!\!\!\rightarrow r)\rightarrow(p\vee q\,\square\!\!\!\rightarrow(\neg p\vee q\rightarrow r))$ Lemma 24 [S]
3. $(p\vee q\,\square\!\!\!\rightarrow(p\vee\neg q\rightarrow r))\rightarrow((p\vee q\,\square\!\!\!\rightarrow(\neg p\vee q\rightarrow r))\rightarrow(p\vee q\,\square\!\!\!\rightarrow r))$ Lemma 25
4. $(p\,\square\!\!\!\rightarrow r)\rightarrow((p\vee q\,\square\!\!\!\rightarrow(\neg p\vee q\rightarrow r))\rightarrow(p\vee q\,\square\!\!\!\rightarrow r))$ 1, 3, DR1
5. $(p\,\square\!\!\!\rightarrow r)\rightarrow((q\,\square\!\!\!\rightarrow r)\rightarrow(p\vee q\,\square\!\!\!\rightarrow r))$ 2, 4, DR3

□

Lemma 23. From S to conclude $(p\,\square\!\!\!\rightarrow r)\rightarrow(p\vee q\,\square\!\!\!\rightarrow(p\vee\neg q\rightarrow r))$.

Proof.

1. $p \to (p \vee q) \wedge (p \vee \neg q)$ pc
2. $(p \vee q) \wedge (p \vee \neg q) \to p$ pc
3. $(p \,\square\!\!\rightarrow r) \to ((p \vee q) \wedge (p \vee \neg q) \,\square\!\!\rightarrow r)$ 1, 2, LLE
4. $(p \wedge q \,\square\!\!\rightarrow r) \to (p \,\square\!\!\rightarrow (q \to r))$ S
5. $((p \vee q) \wedge (p \vee \neg q) \,\square\!\!\rightarrow r) \to (p \vee q \,\square\!\!\rightarrow (p \vee \neg q \to r))$ 4, Sub $(p/p \vee q, q/p \vee \neg q)$
6. $(p \,\square\!\!\rightarrow r) \to (p \vee q \,\square\!\!\rightarrow (p \vee \neg q \to r))$ 3, 5, DR1

$\square$

Lemma 24. From S to conclude $(q \,\square\!\!\rightarrow r) \to (p \vee q \,\square\!\!\rightarrow (\neg p \vee q \to r))$.

Proof.

1. $q \to (p \vee q) \wedge (\neg p \vee q)$ pc
2. $(p \vee q) \wedge (\neg p \vee q) \to q$ pc
3. $(q \,\square\!\!\rightarrow r) \to ((p \vee q) \wedge (\neg p \vee q) \,\square\!\!\rightarrow r)$ 1, 2, LLE
4. $(p \wedge q \,\square\!\!\rightarrow r) \to (p \,\square\!\!\rightarrow (q \to r))$ S
5. $((p \vee q) \wedge (\neg p \vee q) \,\square\!\!\rightarrow r) \to (p \vee q \,\square\!\!\rightarrow (\neg p \vee q \to r))$ 4, Sub $(p/p \vee q, q/\neg p \vee q)$
6. $(q \,\square\!\!\rightarrow r) \to (p \vee q \,\square\!\!\rightarrow (\neg p \vee q \to r))$ 3, 5, DR1

$\square$

Lemma 25. $(p \vee q \,\square\!\!\rightarrow (p \vee \neg q \to r)) \to ((p \vee q \,\square\!\!\rightarrow (\neg p \vee q \to r)) \to (p \vee q \,\square\!\!\rightarrow r))$.

Proof.

1. $(p \,\square\!\!\rightarrow q) \to ((p \,\square\!\!\rightarrow r) \to (p \,\square\!\!\rightarrow q \wedge r))$ AND
2. $(p \vee q \,\square\!\!\rightarrow (p \vee \neg q \to r)) \to ((p \vee q \,\square\!\!\rightarrow (\neg p \vee q \to$ 1, Sub $(p/p \vee q, q/p \vee \neg q \to$
   $r)) \to (p \vee q \,\square\!\!\rightarrow (p \vee \neg q \to r) \wedge (\neg p \vee q \to r)))$ $r, r/\neg p \vee q \to r)$
3. $(p \vee \neg q \to r) \wedge (\neg p \vee q \to r) \to r$ pc
4. $(p \vee q \,\square\!\!\rightarrow (p \vee \neg q \to r) \wedge (\neg p \vee q \to r)) \to (p \vee q \,\square\!\!\rightarrow r)$ 3, RW
5. $(p \vee q \,\square\!\!\rightarrow (p \vee \neg q \to r)) \to ((p \vee q \,\square\!\!\rightarrow (\neg p \vee q \to$ 2, 4, DR2
   $r)) \to (p \vee q \,\square\!\!\rightarrow r))$

$\square$

Lemma 26. From Or and Refl to conclude S (cf. [14, p. 191])

Proof.

1. $p \wedge \neg q \to (q \to r)$ pc
2. $(p \wedge \neg q \,\square\!\!\rightarrow p \wedge \neg q) \to (p \wedge \neg q \,\square\!\!\rightarrow q \to r)$ 1, RW
3. $p \,\square\!\!\rightarrow p$ Refl
4. $p \wedge \neg q \,\square\!\!\rightarrow p \wedge \neg q$ 3, Sub $(p/p \wedge \neg q)$
5. $p \wedge \neg q \,\square\!\!\rightarrow (q \to r)$ 4, 2, MP
6. $(p \,\square\!\!\rightarrow r) \to ((q \,\square\!\!\rightarrow r) \to (p \vee q \,\square\!\!\rightarrow r))$ Or
7. $(p \wedge \neg q \,\square\!\!\rightarrow (q \to r)) \to ((p \wedge q \,\square\!\!\rightarrow (q \to r)) \to$ 6, Sub $(p/p \wedge \neg q, q/p \wedge q, r/q \to$
   $((p \wedge \neg q) \vee (p \wedge q) \,\square\!\!\rightarrow (q \to r)))$ $r)$
8. $(p \wedge q \,\square\!\!\rightarrow (q \to r)) \to ((p \wedge \neg q) \vee (p \wedge q) \,\square\!\!\rightarrow (q \to r))$ 5, 7, MP

9.	$(p \wedge \neg q) \vee (p \wedge q) \to p$	pc
10.	$p \to (p \wedge \neg q) \vee (p \wedge q)$	pc
11.	$((p \wedge \neg q) \vee (p \wedge q) \,\Box\!\!\to (q \to r)) \to (p \,\Box\!\!\to (q \to r))$	9, 10, LLE
12.	$(p \wedge q \,\Box\!\!\to (q \to r)) \to (p \,\Box\!\!\to (q \to r))$	8, 11, DR1
13.	$r \to (q \to r)$	pc
14.	$(p \wedge q \,\Box\!\!\to r) \to (p \wedge q \,\Box\!\!\to (q \to r))$	13, RW
15.	$(p \wedge q \,\Box\!\!\to r) \to (p \,\Box\!\!\to (q \to r))$	14, 12, DR1

$\square$

Let us now prove the equivalence of System **V** and System **R**.

Theorem 27. System **R** is System **V** (cf. [9]; [39, Theorem 7.32]).

Proof. By Lemma 13 RCK*, LE1, LE2, and Refl are rules, respectively theorems, of a logic iff AND, RW, LLE, and Refl are. Let **L** be any logic which renders AND, RW, LLE, and Refl theorems, respective derived rules. To establish the theorem, we have to show that MOD, LV1, and LV2 are in **L** iff CM, Or, and RM are. Lemma 18 implies that MOD, LV1, and LV2 are in **L** iff MOD, LV1, and RM are. Furthermore, due to Lemma 21, CM, Or, and RM are in **L** iff CM, S, and RM are. It remains to be shown that in the presence of RM, (i) CM and S imply MOD and LV1 and (ii) that MOD and LV1 imply CM and S. Lemmata 28 and 32 give us (i), whereas by Lemmata 30 and 33 we obtain (ii). $\square$

Lemma 28. From Refl, CM, and S to conclude MOD.

Proof.

1.	$(\neg p \,\Box\!\!\to p) \to (\neg p \,\Box\!\!\to q)$	Lemma 29 [Refl]
2.	$(p \,\Box\!\!\to r) \to ((p \,\Box\!\!\to q) \to (p \wedge q \,\Box\!\!\to r))$	CM
3.	$(\neg p \,\Box\!\!\to p) \to ((\neg p \,\Box\!\!\to q) \to (\neg p \wedge q \,\Box\!\!\to p))$	2, Sub $(p/\neg p, r/p)$
4.	$(\neg p \,\Box\!\!\to p) \to (\neg p \wedge q \,\Box\!\!\to p)$	1, 3, DR4
5.	$(p \wedge q \,\Box\!\!\to r) \to (p \,\Box\!\!\to (q \to r))$	S
6.	$(q \wedge \neg p \,\Box\!\!\to p) \to (q \,\Box\!\!\to \neg p \to p)$	5, Sub $(p/q, q/\neg p, r/p)$
7.	$\neg p \wedge q \to q \wedge \neg p$	pc
8.	$q \wedge \neg p \to \neg p \wedge q$	pc
9.	$(\neg p \wedge q \,\Box\!\!\to p) \to (q \wedge \neg p \,\Box\!\!\to p)$	7, 8, LLE
10.	$(\neg p \wedge q \,\Box\!\!\to p) \to (q \,\Box\!\!\to (\neg p \to p))$	9, 6, DR1
11.	$(\neg p \,\Box\!\!\to p) \to (q \,\Box\!\!\to (\neg p \to p))$	4, 10, DR1
12.	$(\neg p \to p) \to p$	pc
13.	$(q \,\Box\!\!\to (\neg p \to p)) \to (q \,\Box\!\!\to p)$	12, RW
14.	$(\neg p \,\Box\!\!\to p) \to (q \,\Box\!\!\to p)$	11, 13, DR1

$\square$

Lemma 29. From Refl to conclude $(\neg p\,\square\!\!\rightarrow p) \to (\neg p\,\square\!\!\rightarrow q)$.

Proof.

1. $(p\,\square\!\!\rightarrow q) \to (p\,\square\!\!\rightarrow p \wedge q)$ Refl$^+$ [Refl]
2. $(\neg p\,\square\!\!\rightarrow \neg\neg p) \to (\neg p\,\square\!\!\rightarrow \neg p \wedge \neg\neg p)$ 1, Sub$(p/\neg p, q/\neg\neg p)$
3. $\neg p \wedge \neg\neg p \to q$ pc
4. $(\neg p\,\square\!\!\rightarrow \neg p \wedge \neg\neg p) \to (\neg p\,\square\!\!\rightarrow q)$ 3, RW
5. $(\neg p\,\square\!\!\rightarrow \neg\neg p) \to (\neg p\,\square\!\!\rightarrow q)$ 2, 4, DR1
6. $p \to \neg\neg p$ pc
7. $(\neg p\,\square\!\!\rightarrow p) \to (\neg p\,\square\!\!\rightarrow \neg\neg p)$ 6, RW
8. $(\neg p\,\square\!\!\rightarrow p) \to (\neg p\,\square\!\!\rightarrow q)$ 7, 5, DR1

□

Lemma 30. From Refl, MOD, and RM to conclude CM.

Proof.

1. $(p\,\square\!\!\rightarrow r) \to (\neg(p\,\square\!\!\rightarrow \neg q) \to (p \wedge q\,\square\!\!\rightarrow r))$ RM
2. $(p\,\square\!\!\rightarrow \neg q) \to ((p\,\square\!\!\rightarrow q) \to (p \wedge q\,\square\!\!\rightarrow r))$ Lemma 31 [Refl, MOD]
3. $(p \to (\neg q \to r)) \to ((q \to (s \to r)) \to (p \to (s \to$ pc $r)))$
4. $((p\,\square\!\!\rightarrow r) \to (\neg(p\,\square\!\!\rightarrow \neg q) \to (p \wedge q\,\square\!\!\rightarrow r))) \to$ 3, Sub $(p/p\,\square\!\!\rightarrow r, q/p\,\square\!\!\rightarrow \neg q, r/p \wedge$
   $(((p\,\square\!\!\rightarrow \neg q) \to ((p\,\square\!\!\rightarrow q) \to (p \wedge q\,\square\!\!\rightarrow r))) \to$ $q\,\square\!\!\rightarrow r, s/p\,\square\!\!\rightarrow q)$
   $((p\,\square\!\!\rightarrow r) \to ((p\,\square\!\!\rightarrow q) \to (p \wedge q\,\square\!\!\rightarrow r))))$
5. $((p\,\square\!\!\rightarrow \neg q) \to ((p\,\square\!\!\rightarrow q) \to (p \wedge q\,\square\!\!\rightarrow r))) \to$ 1, 4, MP
   $((p\,\square\!\!\rightarrow r) \to ((p\,\square\!\!\rightarrow q) \to (p \wedge q\,\square\!\!\rightarrow r)))$
6. $(p\,\square\!\!\rightarrow r) \to ((p\,\square\!\!\rightarrow q) \to (p \wedge q\,\square\!\!\rightarrow r))$ 2, 5, MP

□

Lemma 31. From Refl and MOD to conclude $(p\,\square\!\!\rightarrow \neg q) \to ((p\,\square\!\!\rightarrow q) \to (p \wedge q\,\square\!\!\rightarrow r))$.

Proof.

1. $(p\,\square\!\!\rightarrow q) \to ((p\,\square\!\!\rightarrow r) \to (p\,\square\!\!\rightarrow q \wedge r))$ AND
2. $(p\,\square\!\!\rightarrow \neg q) \to ((p\,\square\!\!\rightarrow q) \to (p\,\square\!\!\rightarrow \neg q \wedge q))$ 1, Sub $(q/\neg q, r/q)$
3. $\neg q \wedge q \to \neg p$ pc
4. $(p\,\square\!\!\rightarrow \neg q \wedge q) \to (p\,\square\!\!\rightarrow \neg p)$ 3, RW
5. $p \to \neg\neg p$ pc
6. $\neg\neg p \to p$ pc
7. $(p\,\square\!\!\rightarrow \neg p) \to (\neg\neg p\,\square\!\!\rightarrow \neg p)$ 5, 6, LLE
8. $(p\,\square\!\!\rightarrow \neg q \wedge q) \to (\neg\neg p\,\square\!\!\rightarrow \neg p)$ 4, 7, DR1
9. $(\neg p\,\square\!\!\rightarrow p) \to (q\,\square\!\!\rightarrow p)$ MOD
10. $(\neg\neg p\,\square\!\!\rightarrow \neg p) \to (p \wedge q\,\square\!\!\rightarrow \neg p)$ 9, Sub $(p/\neg p, q/p \wedge q)$
11. $(p\,\square\!\!\rightarrow \neg q \wedge q) \to (p \wedge q\,\square\!\!\rightarrow \neg p)$ 8, 10, DR1
12. $(p\,\square\!\!\rightarrow q) \to (p\,\square\!\!\rightarrow p \wedge q)$ Refl$^+$ [Refl]

13. $(p \wedge q \mathbin{\square\!\!\rightarrow} \neg p) \rightarrow (p \wedge q \mathbin{\square\!\!\rightarrow} p \wedge q \wedge \neg p)$ 12, Sub $(p/p \wedge q, q/\neg p)$

14. $(p \mathbin{\square\!\!\rightarrow} \neg q \wedge q) \rightarrow (p \wedge q \mathbin{\square\!\!\rightarrow} p \wedge q \wedge \neg p)$ 11, 13, DR1

15. $p \wedge q \wedge \neg p \rightarrow r$ pc

16. $(p \wedge q \mathbin{\square\!\!\rightarrow} p \wedge q \wedge \neg p) \rightarrow (p \wedge q \mathbin{\square\!\!\rightarrow} r)$ 15, RW

17. $(p \mathbin{\square\!\!\rightarrow} \neg q \wedge q) \rightarrow (p \wedge q \mathbin{\square\!\!\rightarrow} r)$ 14, 16, DR1

18. $(p \mathbin{\square\!\!\rightarrow} \neg q) \rightarrow ((p \mathbin{\square\!\!\rightarrow} q) \rightarrow (p \wedge q \mathbin{\square\!\!\rightarrow} r))$ 2, 17, DR2

$\square$

Lemma 32. From S to conclude LV1.

Proof. Trivial. $\square$

Lemma 33. From LV1 to conclude S.

Proof.

1. $(p \wedge q \mathbin{\square\!\!\rightarrow} r) \rightarrow (\neg(p \mathbin{\square\!\!\rightarrow} \neg q) \rightarrow (p \mathbin{\square\!\!\rightarrow} (q \rightarrow r)))$ LV1

2. $\neg q \rightarrow (q \rightarrow r)$ pc

3. $(p \mathbin{\square\!\!\rightarrow} \neg q) \rightarrow (p \mathbin{\square\!\!\rightarrow} (q \rightarrow r))$ 2, RW

4. $(p \rightarrow q) \rightarrow ((r \rightarrow (\neg p \rightarrow q)) \rightarrow (r \rightarrow q))$ pc

5. $((p \mathbin{\square\!\!\rightarrow} \neg q) \rightarrow (p \mathbin{\square\!\!\rightarrow} (q \rightarrow r))) \rightarrow (((p \wedge q \mathbin{\square\!\!\rightarrow} r) \rightarrow$ 4, Sub $(p/p \mathbin{\square\!\!\rightarrow} \neg q, q/p \mathbin{\square\!\!\rightarrow}$
$(\neg(p \mathbin{\square\!\!\rightarrow} \neg q) \rightarrow (p \mathbin{\square\!\!\rightarrow} (q \rightarrow r)))) \rightarrow ((p \wedge q \mathbin{\square\!\!\rightarrow} r) \rightarrow$ $(q \rightarrow r), r/p \wedge q \mathbin{\square\!\!\rightarrow} r)$
$(p \mathbin{\square\!\!\rightarrow} (q \rightarrow r))))$

6. $((p \wedge q \mathbin{\square\!\!\rightarrow} r) \rightarrow (\neg(p \mathbin{\square\!\!\rightarrow} \neg q) \rightarrow (p \mathbin{\square\!\!\rightarrow} (q \rightarrow r)))) \rightarrow$ 3, 5, MP
$((p \wedge q \mathbin{\square\!\!\rightarrow} r) \rightarrow (p \mathbin{\square\!\!\rightarrow} (q \rightarrow r)))$

7. $(p \wedge q \mathbin{\square\!\!\rightarrow} r) \rightarrow (p \mathbin{\square\!\!\rightarrow} (q \rightarrow r))$ 1, 6, MP

$\square$

3.2 The Equivalence of Systems $\mathbf{V}^*$ and R

[5] uses the following axioms and rules to axiomatize his system $\mathbf{V}^*$:[16]
Axioms:

Refl $p \mathbin{\square\!\!\rightarrow} p$

AND $(p \mathbin{\square\!\!\rightarrow} q) \rightarrow ((p \mathbin{\square\!\!\rightarrow} r) \rightarrow (p \mathbin{\square\!\!\rightarrow} q \wedge r))$

RW' $(p \mathbin{\square\!\!\rightarrow} q \wedge r) \rightarrow (p \mathbin{\square\!\!\rightarrow} q)$

CM $(p \mathbin{\square\!\!\rightarrow} r) \rightarrow ((p \mathbin{\square\!\!\rightarrow} q) \rightarrow (p \wedge q \mathbin{\square\!\!\rightarrow} r))$

Or $(p \mathbin{\square\!\!\rightarrow} r) \rightarrow ((q \mathbin{\square\!\!\rightarrow} r) \rightarrow (p \vee q \mathbin{\square\!\!\rightarrow} r))$

RM* $(p \vee r \mathbin{\square\!\!\rightarrow} \neg r) \rightarrow (\neg(p \vee q \mathbin{\square\!\!\rightarrow} \neg q) \rightarrow (q \vee r \mathbin{\square\!\!\rightarrow} \neg r))$

---

[16] [5] calls RM* 'D'' (p. 82). RM* is logically equivalent to $D'[p/q, q/r, r/p]$ in classical conditional logics.

Rules:

LE Exchange of Logical Equivalents:

    LE1 from $\alpha \to \beta$ and $\beta \to \alpha$ to conclude $(\alpha \,\square\!\!\to \gamma) \to (\beta \,\square\!\!\to \gamma)$

    LE2 from $\beta \to \gamma$ and $\gamma \to \beta$ to conclude $(\alpha \,\square\!\!\to \beta) \to (\alpha \,\square\!\!\to \gamma)$

It is not difficult to see that the following lemma holds:

Lemma 34. From AND, RW′, LE1, LE2, and Refl to conclude AND, RW, LLE, and Refl, and vice versa.

Proof. LE1 is nothing but LLE. LE2 follows trivially from RW. It remains to be shown that (i) that RW′ is a theorem of AND, RW, LLE, and Refl and (ii) that RW is derivable from AND, RW′, LE1, LE2, and Refl. (i) holds trivially and Lemma 35 ensures that (ii) holds as well. $\square$

Lemma 35. RW is derivable if RW′ is a theorem and LE2 is derivable.

Proof.

1. $\beta \to \gamma$	Condition of RW	
2. $(\beta \to \gamma) \to (\beta \to \gamma \wedge \beta)$	pc	
3. $\beta \to \gamma \wedge \beta$	+1	1, 2, MP
4. $\gamma \wedge \beta \to \gamma$	pc	
5. $(\alpha \,\square\!\!\to \beta) \to (\alpha \,\square\!\!\to \gamma \wedge \beta)$	+1	3, 4, LE2
6. $(p \,\square\!\!\to q \wedge r) \to (p \,\square\!\!\to q)$	RW′	
7. $(\alpha \,\square\!\!\to \gamma \wedge \beta) \to (\alpha \,\square\!\!\to \gamma)$	6, Sub $(p/\alpha, q/\gamma, r/\beta)$	
8. $(\alpha \,\square\!\!\to \beta) \to (\alpha \,\square\!\!\to \gamma)$	+1	5, 7, DR1

$\square$

Corollary 36. System $\mathbf{V}^*$ is a classical conditional logic.

Let us now state the equivalence theorem for System $\mathbf{V}^*$ and System $\mathbf{R}$.

Theorem 37. System $\mathbf{V}^*$ is System $\mathbf{R}$.

Proof. Lemma 34 shows that AND, RW′, LE1, LE2, and Refl are equivalent to AND, RW, LLE, and Refl. Thus, it remains to be proved that (i) RM* is a theorem of System $\mathbf{R}$ and that (ii) RM is a theorem of System $\mathbf{V}^*$. Lemmata 38 and 42 yield (i) and (ii), respectively, and give us the desired result. $\square$

Lemma 38. From Refl, CM, Or, and RM to conclude RM*.

Proof.

1. $(p \lor q \lor r \,\square\!\!\rightarrow\, \neg r) \rightarrow (\neg(p \lor q \lor r \,\square\!\!\rightarrow\, \neg(\neg p$ Lemma 39 [RM]
   $\lor\, q \lor r)) \rightarrow (q \lor r \,\square\!\!\rightarrow\, \neg r))$

2. $(p \lor r \,\square\!\!\rightarrow\, \neg r) \rightarrow (p \lor q \lor r \,\square\!\!\rightarrow\, \neg r)$ Lemma 40 [Refl, Or]

3. $(p \lor q \lor r \,\square\!\!\rightarrow\, \neg r) \rightarrow ((p \lor q \lor r \,\square\!\!\rightarrow\, \neg(\neg p$ Lemma 41 [CM]
   $\lor\, q \lor r)) \rightarrow (p \lor q \,\square\!\!\rightarrow\, \neg q))$

4. $(p \lor q \lor r \,\square\!\!\rightarrow\, \neg r) \rightarrow (\neg(p \lor q \,\square\!\!\rightarrow\, \neg q) \rightarrow$ 1, 3, DR6 $(p/p \lor q \lor r \,\square\!\!\rightarrow\, \neg r,\ q/p \lor q \lor r \,\square\!\!\rightarrow\,$
   $(q \lor r \,\square\!\!\rightarrow\, \neg r))$ $\neg(\neg p \lor q \lor r),\ r/q \lor r \,\square\!\!\rightarrow\, \neg r,\ s/p \lor q \,\square\!\!\rightarrow\, \neg q)$

5. $(p \lor r \,\square\!\!\rightarrow\, \neg r) \rightarrow (\neg(p \lor q \,\square\!\!\rightarrow\, \neg q) \rightarrow$ 2, 4, DR1
   $(q \lor r \,\square\!\!\rightarrow\, \neg r))$

$\square$

Lemma 39. From RM to conclude $(p \lor q \lor r \,\square\!\!\rightarrow\, \neg r) \rightarrow (\neg(p \lor q \lor r \,\square\!\!\rightarrow\, \neg(\neg p \lor q \lor r)) \rightarrow (q \lor r \,\square\!\!\rightarrow\, \neg r))$.

Proof.

1. $(p \,\square\!\!\rightarrow\, r) \rightarrow (\neg(p \,\square\!\!\rightarrow\, \neg q) \rightarrow (p \land q \,\square\!\!\rightarrow\, r))$ RM

2. $(p \lor q \lor r \,\square\!\!\rightarrow\, \neg r) \rightarrow (\neg(p \lor q \lor r \,\square\!\!\rightarrow\, \neg(\neg p \lor q \lor r)) \rightarrow$ 1, Sub $(p/p \lor q \lor r,$
   $((p \lor q \lor r) \land (\neg p \lor q \lor r) \,\square\!\!\rightarrow\, \neg r)))$ $q/\neg p \lor q \lor r,\ r/\neg r)$

3. $(p \lor q \lor r) \land (\neg p \lor q \lor r) \rightarrow q \lor r$ pc

4. $q \lor r \rightarrow (p \lor q \lor r) \land (\neg p \lor q \lor r)$ pc

5. $((p \lor q \lor r) \land (\neg p \lor q \lor r) \,\square\!\!\rightarrow\, \neg r) \rightarrow (q \lor r \,\square\!\!\rightarrow\, \neg r)$ 3, 4, LLE

6. $(p \lor q \lor r \,\square\!\!\rightarrow\, \neg r) \rightarrow (\neg(p \lor q \lor r \,\square\!\!\rightarrow\, \neg(\neg p \lor q \lor r)) \rightarrow (q \lor r \,\square\!\!\rightarrow\, \neg r))$ 2, 5, DR2

$\square$

Lemma 40. From Refl and Or to conclude $(p \lor r \,\square\!\!\rightarrow\, \neg r) \rightarrow (p \lor q \lor r \,\square\!\!\rightarrow\, \neg r)$.

Proof.

1. $p \lor r \rightarrow (p \lor q \lor r) \land (p \lor \neg q \lor r)$ pc

2. $(p \lor q \lor r) \land (p \lor \neg q \lor r) \rightarrow p \lor r$ pc

3. $(p \lor r \,\square\!\!\rightarrow\, \neg r) \rightarrow ((p \lor q \lor r) \land (p \lor \neg q \lor r) \,\square\!\!\rightarrow\, \neg r)$ 1, 2, LLE

4. $(p \land q \,\square\!\!\rightarrow\, r) \rightarrow (p \,\square\!\!\rightarrow\, (q \rightarrow r))$ Lemma 21 [Refl, Or]

5. $((p \lor q \lor r) \land (p \lor \neg q \lor r) \,\square\!\!\rightarrow\, \neg r) \rightarrow (p \lor q \lor r \,\square\!\!\rightarrow\,$ 4, Sub $(p/p \lor q \lor r,\ q/p \lor \neg q$
   $(p \lor \neg q \lor r \rightarrow \neg r))$ $\lor\, r,\ r/\neg r)$

6. $(p \lor r \,\square\!\!\rightarrow\, \neg r) \rightarrow (p \lor q \lor r \,\square\!\!\rightarrow\, (p \lor \neg q \lor r \rightarrow \neg r))$ 3, 5, DR1

7. $(p \lor \neg q \lor r \rightarrow \neg r) \rightarrow \neg r$ pc

8. $(p \lor q \lor r \,\square\!\!\rightarrow\, (p \lor \neg q \lor r \rightarrow \neg r)) \rightarrow (p \lor q \lor r \,\square\!\!\rightarrow\, \neg r)$ 7, RW

9. $(p \lor r \,\square\!\!\rightarrow\, \neg r) \rightarrow (p \lor q \lor r \,\square\!\!\rightarrow\, \neg r)$ 6, 8, DR1

$\square$

Lemma 41. From CM to conclude $(p \lor q \lor r \,\square\!\!\rightarrow\, \neg r) \to ((p \lor q \lor r \,\square\!\!\rightarrow\, \neg(\neg p \lor q \lor r)) \to (p \lor q \,\square\!\!\rightarrow\, \neg q))$.

Proof.

1.	$(p \,\square\!\!\rightarrow\, r) \to ((p \,\square\!\!\rightarrow\, q) \to (p \land q \,\square\!\!\rightarrow\, r))$	CM
2.	$(p \lor q \lor r \,\square\!\!\rightarrow\, \neg q) \to ((p \lor q \lor r \,\square\!\!\rightarrow\, p \lor q \lor \neg r) \to$ $((p \lor q \lor r) \land (p \lor q \lor \neg r) \,\square\!\!\rightarrow\, \neg q))$	1, Sub $(p/p \lor q \lor r, q/p \lor q \lor \neg r, r/\neg q)$
3.	$(p \lor q \lor r) \land (p \lor q \lor \neg r) \to p \lor q$	pc
4.	$p \lor q \to (p \lor q \lor r) \land (p \lor q \lor \neg r)$	pc
5.	$((p \lor q \lor r) \land (p \lor q \lor \neg r) \,\square\!\!\rightarrow\, \neg q) \to (p \lor q \,\square\!\!\rightarrow\, \neg q)$	3, 4, LLE
6.	$(p \lor q \lor r \,\square\!\!\rightarrow\, \neg q) \to ((p \lor q \lor r \,\square\!\!\rightarrow\, p \lor q \lor \neg r) \to (p \lor q \,\square\!\!\rightarrow\, \neg q))$	2, 5, DR2
7.	$\neg(\neg p \lor q \lor r) \to \neg q$	pc
8.	$(p \lor q \lor r \,\square\!\!\rightarrow\, \neg(\neg p \lor q \lor r)) \to (p \lor q \lor r \,\square\!\!\rightarrow\, \neg q)$	7, RW
9.	$(p \lor q \lor r \,\square\!\!\rightarrow\, \neg(\neg p \lor q \lor r)) \to$ $((p \lor q \lor r \,\square\!\!\rightarrow\, p \lor q \lor \neg r) \to (p \lor q \,\square\!\!\rightarrow\, \neg q))$	8, 6, DR1
10.	$\neg r \to p \lor q \lor \neg r$	pc
11.	$(p \lor q \lor r \,\square\!\!\rightarrow\, \neg r) \to (p \lor q \lor r \,\square\!\!\rightarrow\, p \lor q \lor \neg r)$	10, RW
12.	$(p \lor q \lor r \,\square\!\!\rightarrow\, \neg r) \to ((p \lor q \lor r \,\square\!\!\rightarrow\, \neg(\neg p \lor q \lor r)) \to (p \lor q \,\square\!\!\rightarrow\, \neg q))$	11, 9, DR5 $(p/p \lor q \lor r \,\square\!\!\rightarrow\, \neg r, q/p \lor q \lor r \,\square\!\!\rightarrow\, p \lor q \lor \neg r, r/p \lor q \lor r \,\square\!\!\rightarrow\, \neg(\neg p \lor q \lor r), s/p \lor q \,\square\!\!\rightarrow\, \neg q)$

$\square$

Lemma 42. From Refl and RM* to conclude RM.

Proof.

1.	$(p \lor (p \land q \land \neg r) \,\square\!\!\rightarrow\, \neg(p \land q \land \neg r)) \to (\neg(p \lor (p \land q \land r) \,\square\!\!\rightarrow\, \neg(p \land q \land r)) \to (p \land q \,\square\!\!\rightarrow\, r))$	Lemma 43 [Refl, RM*]
2.	$(p \,\square\!\!\rightarrow\, r) \to (p \lor (p \land q \land \neg r) \,\square\!\!\rightarrow\, \neg(p \land q \land \neg r))$	Lemma 44
3.	$(p \,\square\!\!\rightarrow\, r) \to ((p \lor (p \land q \land r) \,\square\!\!\rightarrow\, \neg(p \land q \land r)) \to (p \,\square\!\!\rightarrow\, \neg q))$	Lemma 45 [Refl]
4.	$(p \,\square\!\!\rightarrow\, r) \to (\neg(p \lor (p \land q \land r) \,\square\!\!\rightarrow\, \neg(p \land q \land r)) \to (p \land q \,\square\!\!\rightarrow\, r))$	2, 1, DR1
5.	$(p \,\square\!\!\rightarrow\, r) \to (\neg(p \,\square\!\!\rightarrow\, \neg q) \to (p \land q \,\square\!\!\rightarrow\, r))$	4, 3, DR6 $(p/p \,\square\!\!\rightarrow\, r, q/p \lor (p \land q \land r) \,\square\!\!\rightarrow\, \neg(p \land q \land r), r/p \land q \,\square\!\!\rightarrow\, r, s/p \,\square\!\!\rightarrow\, \neg q)$

$\square$

Lemma 43. From Refl and RM* to conclude $(p \lor (p \land q \land \neg r) \,\square\!\!\rightarrow\, \neg(p \land q \land \neg r)) \to (\neg(p \lor (p \land q \land r) \,\square\!\!\rightarrow\, \neg(p \land q \land r)) \to (p \land q \,\square\!\!\rightarrow\, r))$.

Proof.
1. $(p \vee r \,\square\!\!\rightarrow \neg r) \rightarrow (\neg(p \vee q \,\square\!\!\rightarrow \neg q) \rightarrow (q \vee r \,\square\!\!\rightarrow \neg r))$ RM*
2. $(p \vee (p \wedge q \wedge \neg r) \,\square\!\!\rightarrow \neg(p \wedge q \wedge \neg r)) \rightarrow ((\neg(p \vee (p \wedge q \wedge r) \,\square\!\!\rightarrow$ 1, Sub $(q/p \wedge q \wedge r, r/p \wedge$
   $\neg(p \wedge q \wedge r)) \rightarrow ((p \wedge q \wedge r) \vee (p \wedge q \wedge \neg r) \,\square\!\!\rightarrow \neg(p \wedge q \wedge \neg r)))$ $q \wedge \neg r)$
3. $(p \wedge q \wedge r) \vee (p \wedge q \wedge \neg r) \rightarrow p \wedge q$ pc
4. $p \wedge q \rightarrow (p \wedge q \wedge r) \vee (p \wedge q \wedge \neg r)$ pc
5. $((p \wedge q \wedge r) \vee (p \wedge q \wedge \neg r) \,\square\!\!\rightarrow \neg(p \wedge q \wedge \neg r)) \rightarrow (p \wedge q \,\square\!\!\rightarrow$ 3, 4, LLE
   $\neg(p \wedge q \wedge \neg r))$
6. $(p \vee (p \wedge q \wedge \neg r) \,\square\!\!\rightarrow \neg(p \wedge q \wedge \neg r)) \rightarrow (\neg(p \vee (p \wedge q \wedge r) \,\square\!\!\rightarrow$ 2, 5, DR2
   $\neg(p \wedge q \wedge r)) \rightarrow (p \wedge q \,\square\!\!\rightarrow \neg(p \wedge q \wedge \neg r)))$
7. $(p \,\square\!\!\rightarrow q) \rightarrow (p \,\square\!\!\rightarrow p \wedge q)$ Refl$^+$ [Refl]
8. $(p \wedge q \,\square\!\!\rightarrow \neg(p \wedge q \wedge \neg r)) \rightarrow (p \wedge q \,\square\!\!\rightarrow p \wedge q \wedge \neg(p \wedge q \wedge \neg r))$ 7, Sub $(p/p \wedge q, q/\neg(p \wedge$
   $q \wedge \neg r))$
9. $(p \vee (p \wedge q \wedge \neg r) \,\square\!\!\rightarrow \neg(p \wedge q \wedge \neg r)) \rightarrow (\neg(p \vee (p \wedge q \wedge r) \,\square\!\!\rightarrow$ 6, 8, DR2
   $\neg(p \wedge q \wedge r)) \rightarrow (p \wedge q \,\square\!\!\rightarrow p \wedge q \wedge \neg(p \wedge q \wedge \neg r)))$
10. $p \wedge q \wedge \neg(p \wedge q \wedge \neg r) \rightarrow r$ pc
11. $(p \wedge q \,\square\!\!\rightarrow p \wedge q \wedge \neg(p \wedge q \wedge \neg r)) \rightarrow (p \wedge q \,\square\!\!\rightarrow r)$ 10, RW
12. $(p \vee (p \wedge q \wedge \neg r) \,\square\!\!\rightarrow \neg(p \wedge q \wedge \neg r)) \rightarrow (\neg(p \vee (p \wedge q \wedge r) \,\square\!\!\rightarrow$ 9, 11, DR2
    $\neg(p \wedge q \wedge r)) \rightarrow (p \wedge q \,\square\!\!\rightarrow r))$

$\square$

Lemma 44. $(p \,\square\!\!\rightarrow r) \rightarrow (p \vee (p \wedge q \wedge \neg r) \,\square\!\!\rightarrow \neg(p \wedge q \wedge \neg r))$

Proof.
1. $p \rightarrow p \vee (p \wedge q \wedge \neg r)$ pc
2. $p \vee (p \wedge q \wedge \neg r) \rightarrow p$ pc
3. $(p \,\square\!\!\rightarrow r) \rightarrow (p \vee (p \wedge q \wedge \neg r) \,\square\!\!\rightarrow r)$ 1, 2, LLE
4. $r \rightarrow \neg(p \wedge q \wedge \neg r)$ pc
5. $(p \vee (p \wedge q \wedge \neg r) \,\square\!\!\rightarrow r) \rightarrow (p \vee (p \wedge q \wedge \neg r) \,\square\!\!\rightarrow$ 4, RW
   $\neg(p \wedge q \wedge \neg r))$
6. $(p \,\square\!\!\rightarrow r) \rightarrow (p \vee (p \wedge q \wedge \neg r) \,\square\!\!\rightarrow \neg(p \wedge q \wedge \neg r))$ 3, 5, DR1

$\square$

Lemma 45. From Refl to conclude $(p \,\square\!\!\rightarrow r) \rightarrow ((p \vee (p \wedge q \wedge r) \,\square\!\!\rightarrow \neg(p \wedge q \wedge r)) \rightarrow (p \,\square\!\!\rightarrow \neg q))$.

Proof.
1. $(p \,\square\!\!\rightarrow q) \rightarrow ((p \,\square\!\!\rightarrow r) \rightarrow (p \,\square\!\!\rightarrow q \wedge r))$ AND
2. $(p \,\square\!\!\rightarrow r) \rightarrow ((p \,\square\!\!\rightarrow \neg(p \wedge q \wedge r)) \rightarrow (p \,\square\!\!\rightarrow r \wedge \neg(p \wedge$ 1, Sub $(q/r, r/\neg(p \wedge q \wedge r))$
   $q \wedge r)))$
3. $(p \,\square\!\!\rightarrow q) \rightarrow (p \,\square\!\!\rightarrow p \wedge q)$ Refl$^+$ [Refl]

4. $(p\Box\!\!\!\rightarrow r\wedge\neg(p\wedge q\wedge r))\rightarrow(p\Box\!\!\!\rightarrow p\wedge r\wedge\neg(p\wedge q\wedge r))$ 3, Sub $(q/r\wedge\neg(p\wedge q\wedge r))$
5. $(p\Box\!\!\!\rightarrow r)\rightarrow((p\Box\!\!\!\rightarrow\neg(p\wedge q\wedge r))\rightarrow(p\Box\!\!\!\rightarrow p\wedge r\wedge$ 2, 4, DR2
    $\neg(p\wedge q\wedge r)))$
6. $p\wedge r\wedge\neg(p\wedge q\wedge r)\rightarrow\neg q$ pc
7. $(p\Box\!\!\!\rightarrow p\wedge r\wedge\neg(p\wedge q\wedge r))\rightarrow(p\Box\!\!\!\rightarrow\neg q)$ 6, RW
8. $(p\Box\!\!\!\rightarrow r)\rightarrow((p\Box\!\!\!\rightarrow\neg(p\wedge q\wedge r))\rightarrow(p\Box\!\!\!\rightarrow\neg q))$ 5, 7, DR2
9. $p\vee(p\wedge q\wedge r)\rightarrow p$ pc
10. $p\rightarrow p\vee(p\wedge q\wedge r)$ pc
11. $(p\vee(p\wedge q\wedge r)\Box\!\!\!\rightarrow\neg(p\wedge q\wedge r))\rightarrow(p\Box\!\!\!\rightarrow\neg(p\wedge q\wedge r))$ 9, 10, LLE
12. $(p\Box\!\!\!\rightarrow r)\rightarrow((p\vee(p\wedge q\wedge r)\Box\!\!\!\rightarrow\neg(p\wedge q\wedge r))\rightarrow$ 11, 8, DR3 $(p/p\vee(p\wedge q\wedge r)\Box\!\!\!\rightarrow$
    $(p\Box\!\!\!\rightarrow\neg q))$ $\neg(p\wedge q\wedge r),\ q/p\Box\!\!\!\rightarrow\neg(p\wedge q\wedge r),$
    $r/p\Box\!\!\!\rightarrow r,\ s/p\Box\!\!\!\rightarrow\neg q)$

$\Box$

4 Connexive Principles and the Relation between Cautious Monotonicity and Rational Monotonicity

This section shows which role default connexive principles, DAT and DBT, play in the axiomatization of Systems **P** and **R**. By the equivalence of Systems **R**, **V**, and **V*** this extends also to the latter two systems. I shall show that CM is a theorem rather than an axiom of the axiomatization of System **R** in the presence of DAT and DBT, whereas it is not derivable from System **P** when DAT or an equivalent connexive principle is missing.

To obtain the former result, I shall extend the proof system described in Section 1 by the new rule TNT ('Theorem Non-Theorem'). TNT is analogous to a proof by cases as employed in natural deduction systems. Proofs by cases license inferences of a formula β simpliciter – i.e., without either assumption α or $\neg\alpha$ – if β is derivable from both α and from $\neg\alpha$. Now TNT asserts that a formula β is provable simpliciter – without reference to either a provability or a non-provability assumption – if the same formula β is derivable both from the assumption that a formula α is provable (theoremhood) and from the assumption that α is not provable (non-theoremhood). TNT differs from a traditional proof by cases only in that TNT refers to the provability/non-provability of a formula where a traditional proof by cases refers to the assumption that a formula and its negation hold.

To formulate TNT, I use a notation akin to proof by cases in Fitch-style natural deduction systems. By TNT also sequences of formulae as follows count as proofs, where lines (m+1) and (n+1) are provability (theoremhood) and non-provability (non-theoremhood) assumptions, respectively:

$$m. \boxed{\vdash} \alpha \qquad \text{Theorem}$$

$$\vdots$$

$$n. \beta \qquad +m$$

$$(n+1). \boxed{\nvdash} \alpha \qquad \text{Non-Theorem}$$

$$\vdots$$

$$o. \beta \qquad -(n+1)$$

$$(o+1). \quad \beta \qquad \text{(m–n), (n+1)–o, TNT}$$

The comment '+m' in line n indicates that the derivation of β depends on the provability of α as described in line m, whereas '$-(n+1)$' in line o indicates that the derivation of β depends on the non-provability of α, as stated in line $(n+1)$. The successful application of TNT includes the introduction of provability and non-provability assumption as described by lines m and $(n+1)$. TNT allows then to dispense with assumptions $+m$ and $-(n+1)$ and to infer β without provability or non-provability assumption. In addition, a reiteration rule (Reit) is required that allows to repeat lines from subordinate (sub)proofs to superordinate (sub)proofs. In the above schema m–n is assumed to be a superordinate proof of a (sub)proof that includes lines 1–(o+1). An example of an application of Reit is line 15 in the proof of Lemma 48.

Let me now show that System **R** has CM as a theorem rather than an axiom when **R**, **V**, or **V*** is augmented by either DBT or DAT.

Theorem 46. CM is derivable from the classical conditional logics which have the following principles as theorems or rules:

(1) Refl, DBT, and RM (3) Refl, DBT, and RM* (5) Refl, DBT, and LV2
(2) Refl, DAT, and RM (4) Refl, DAT, and RM* (6) Refl, DAT, and LV2

Proof. (1) follows from Lemma 48. Based on (1) Lemmata 42 and 19 imply (3) and (5), respectively. Finally, Lemma 7 implies that (2), (4), and (6) follow from (1), (3), and (5), respectively. □

Corollary 47. System **R** has CM as a theorem rather than an axiom when **R** is augmented by either principles DAT or DBT.

Lemma 48. From Refl, RM, and DBT to conclude CM.
Proof. Let me comment on the proof below. The assumption $\neg\alpha$ in line 3 states that $\neg\alpha$ is provable (theoremhood), whereas the assumption $\neg\alpha$ in line 13 states

that $\neg\alpha$ is not provable (non-theoremhood). The Rule TNT lets us infer line 17 from lines (3)–(12) and (13)–(16), since the formula in line 17 is derivable from both the provability assumption in line 3 and the non-provability assumption in line 13. Finally, principle DBT can be applied in line 14, since α being consistent is equivalent to $\neg\alpha$ not being provable (see Sections 1 and 2).

1.	$(p\,\Box\!\!\rightarrow r)\rightarrow(\neg(p\,\Box\!\!\rightarrow\neg q)\rightarrow(p\wedge q\,\Box\!\!\rightarrow r))$	RM
2.	$(\alpha\,\Box\!\!\rightarrow\gamma)\rightarrow(\neg(\alpha\,\Box\!\!\rightarrow\neg\beta)\rightarrow(\alpha\wedge\beta\,\Box\!\!\rightarrow\gamma))$	1, Sub $(p/\alpha,q/\beta,r/\gamma)$
3.	$\vdash \neg\alpha$	Theorem
4.	$\neg\alpha\rightarrow(\alpha\wedge\beta\rightarrow\gamma)$	pc
5.	$\alpha\wedge\beta\rightarrow\gamma$	+3 3, 4, MP
6.	$(\alpha\wedge\beta\,\Box\!\!\rightarrow\alpha\wedge\beta)\rightarrow(\alpha\wedge\beta\,\Box\!\!\rightarrow\gamma)$	+3 5, RW
7.	$(p\,\Box\!\!\rightarrow p)$	Refl
8.	$(\alpha\wedge\beta\,\Box\!\!\rightarrow\alpha\wedge\beta)$	7, Sub $(p/\alpha\wedge\beta)$
9.	$(\alpha\wedge\beta\,\Box\!\!\rightarrow\gamma)$	+3 8, 6, MP
10.	$p\rightarrow(q\rightarrow(r\rightarrow p))$	pc
11.	$(\alpha\wedge\beta\,\Box\!\!\rightarrow\gamma)\rightarrow((\alpha\,\Box\!\!\rightarrow\gamma)\rightarrow((\alpha\,\Box\!\!\rightarrow\beta)\rightarrow$ $(\alpha\wedge\beta\,\Box\!\!\rightarrow\gamma)))$	10, Sub $(p/\alpha\wedge\beta\,\Box\!\!\rightarrow\gamma,q/\alpha\,\Box\!\!\rightarrow$ $\gamma,r/\alpha\,\Box\!\!\rightarrow\beta)$
12.	$(\alpha\,\Box\!\!\rightarrow\gamma)\rightarrow((\alpha\,\Box\!\!\rightarrow\beta)\rightarrow(\alpha\wedge\beta\,\Box\!\!\rightarrow\gamma))$	+3 9, 11, MP
13.	$\nvdash \neg\alpha$	Non-Theorem
14.	$(\alpha\,\Box\!\!\rightarrow\beta)\rightarrow\neg(\alpha\,\Box\!\!\rightarrow\neg\beta)$	−13 13, DBT
15.	$(\alpha\,\Box\!\!\rightarrow\gamma)\rightarrow(\neg(\alpha\,\Box\!\!\rightarrow\neg\beta)\rightarrow(\alpha\wedge\beta\,\Box\!\!\rightarrow\gamma))$	2, Reit
16.	$(\alpha\,\Box\!\!\rightarrow\gamma)\rightarrow((\alpha\,\Box\!\!\rightarrow\beta)\rightarrow(\alpha\wedge\beta\,\Box\!\!\rightarrow\gamma))$	−13 14, 15, DR4
17.	$(\alpha\,\Box\!\!\rightarrow\gamma)\rightarrow((\alpha\,\Box\!\!\rightarrow\beta)\rightarrow(\alpha\wedge\beta\,\Box\!\!\rightarrow\gamma))$	3–12, 13–16, TNT

$\square$

Let me, finally, show that CM does not follow from the remaining axioms of System **P**. To strengthen this result I shall show that CM neither follows when the following principle is assumed to be a theorem of System **P** ('CC' for 'Cautious Cut'):

$$(p\wedge q\,\Box\!\!\rightarrow r)\rightarrow((p\,\Box\!\!\rightarrow q)\rightarrow(p\,\Box\!\!\rightarrow r))\ \ (CC)$$

The frame conditions that correspond to the Refl, CC, CM, Or, and RM are as follows ([39, Ch. 5]; [40, p. 905]):

Refl $\forall X\forall w,w'(wR_Xw'\Rightarrow w'\in X)$

CC $\forall X,Y\forall w(\forall w'(wR_Xw'\Rightarrow w'\in Y)\Rightarrow\forall w'(wR_Xw'\Rightarrow wR_{X\cap Y}w'))$

CM $\forall X,Y\forall w(\forall w'(wR_Xw'\Rightarrow w'\in Y)\Rightarrow\forall w'(wR_{X\cap Y}w'\Rightarrow wR_Xw'))$

Or $\forall X,Y\forall w,w'\ (wR_{X\cup Y}w'\Rightarrow wR_Xw'\curlyvee wR_Yw')$

RM $\forall X,Y\forall w(\exists w'(wR_Xw'\curlywedge w'\in Y)\Rightarrow\forall w'(wR_{X\cap Y}w'\Rightarrow wR_Xw'))$

Lemma 49. From Refl, CC, Or, and RM not to conclude CM.

Proof. We show that there exists a Chellas frame $\langle W, R \rangle$ that satisfies C_{Refl}, C_{CC}, C_{Or}, as well as C_{RM} and does not satisfy C_{CM}. By the correspondence results for C_{Refl}, C_{RM}, C_{CC}, C_{Or}, and C_{CM} any theorem of a classical logic that contains only Refl, RM, CC, and Or is valid on $\langle W, R \rangle$. However, CM is not valid on $\langle W, R \rangle$, as its corresponding structural condition is not satisfied. Thus, the extension of a classical conditional logic by Refl, CC, Or, and RM does not suffice to render CM a theorem.

Let us show that such a frame exists. Let $\langle W, R \rangle$ be a Chellas frame such that $W = \{w, w'\}$ and $R = \{\langle w, w', \{w'\}\rangle\}$. It is easy to see that $\langle W, R \rangle$ satisfies C_{Refl}, C_{RM}, C_{CC}, and C_{Or}. However, C_{CM} does not hold for $\langle W, R \rangle$, since for $X = W$ and $Y = \{w'\}$ it is not the case that $wR_X w'$ although $\forall w''(wR_X w'' \Rightarrow w'' \in Y)$ and $wR_{X \cap Y} w'$ are satisfied. □

5 Conclusion

The results of the present paper strongly suggest connexive principles warrant further study. They are at the center stage of conditional logics, such as Systems **P**, **R**, **V**, and **V***. At the same time their role in conditional logics is far from fully explored.

On the other hand, the present paper gives a diagnosis of the problem, based on the strengthened impossibility result for connexive principles in classical conditional logics. It is shown that either AT and BT or Refl and LT, respectively, have to be restricted, to arrive at a consistent and complete (classical) conditional logic. Such a restriction can than either be achieved in terms of (i) consistency requirements or (ii) possibility conditions. This yields four different avenues for consistent and complete (classical) conditional logics, as described by (a) DAT and DBT, (b) DRefl and DLT, (c) PAT and PBT, or (d) PRefl and PLT, where (c) has been seen to trivialize AT and BT (see the introductory section).

In this paper we focused on the approach (a), which is, for example, endorsed by [3] and [22]. On such an account, connexive principles are far from inconsequential, insofar in the presence of either DAT or DBT, RM implies CM, thereby introducing a proof-theoretic dependency between two key principles of conditional logics. To give a full account of the proof theory of such systems we do not only have to allow for both theoremhood and non-theoremhood conditions – as it is generally recognized – but also proof theoretic principles such as TNT described in Section 4.

Let me comment on the alternative approaches (b) and (d). Approach (b) seems certainly worthwhile exploring. Taking (b) rather than (a) gives precedence of connexive principles over Refl and and LT, which are arguably also core principles of conditional logics. Such an approach might be less plausible with DLT as the resulting logic is not strictly speaking a classical conditional logic. As an alternative we might use classical conditional logics which do not require the set of consequents to include logical truths (e.g., [6, Sect. 8], see introductory section) or otherwise use a minimal conditional logic with a base logic other than classical logic.

This leaves approach (d). Both [29] and [20, p. Sect. 1.6] endorse such an approach, as I shall argue. [29] and [20, p. Sect. 1.6] define truth conditions for connexive conditional $\alpha \boxdot\!\!\Rightarrow \beta$ based on truth conditions for non-connexive conditionals $\alpha \boxdot\!\!\rightarrow \beta$ as described in the previous sections. To achieve this, the authors require for $\alpha \boxdot\!\!\Rightarrow \beta$ to be true that $\Diamond\alpha$ ("α is possible") is true and $\alpha \boxdot\!\!\rightarrow \beta$ is true. While this approach seems to cash out an intuitively correct set of models, it weakens the proof theory that characterizes $\alpha \boxdot\!\!\Rightarrow \beta$. Firstly, note that for $\alpha \boxdot\!\!\rightarrow \beta$ so defined $\alpha \boxdot\!\!\Rightarrow \alpha$ is only valid if $\Diamond\alpha$ is valid. Thus, by the use of $\Diamond\alpha$, Refl is effectively restricted to PRefl above. Secondly, whereas the set of models described by the connexive conditional $\alpha \boxdot\!\!\Rightarrow \beta$ seems intuitively valid, it effects a logic which trivializes the application of Refl. The only formulae for which $\Diamond\alpha$ is guaranteed to be valid are logically true formulae α. As a consequence, the only formulae for which $\alpha \boxdot\!\!\Rightarrow \alpha$ is valid in all models are logically true formulae α.

In contrast, we might not need to restrict either of the above principles if we employ a non-classical conditional logic. In fact, such an approach is the dominant one in the literature on connexive logics [43]. In this literature it is an open question how to characterize conditional logic systems, such as **P** and **R**, when they are augmented by full connexive principles (cf. [23]; see also [28]).

References

[1] E. Adams. The logic of conditionals. *Inquiry*, 8(1–4):166–197, 1965.

[2] E. Adams. Probability and the logic of conditionals. In J. Hintikka and P. Suppes, editors, *Aspects of Inductive Logic*, pages 265–316. North-Holland Publishing Company, Amsterdam, 1966.

[3] E. Adams. *The Logic of Conditionals*. D. Reidel, Dordrecht, Netherlands, 1975.

[4] E. Adams. On the logic of high probability. *Journal of Philosophical Logic*, 15(3):255–279, 1986.

[5] J. P. Burgess. Quick completeness proofs for some logics of conditionals. *Notre Dame Journal of Formal Logic*, 22(1):76–84, 1981.

[6] B. F. Chellas. Basic conditional logic. *Journal of Philosophical Logic*, 4(2):133–153, 1975.

[7] I. Douven. *The Epistemology of Indicative Conditionals: Formal and Empirical Approaches*. Cambridge University Press, Cambridge, 2015.

[8] H. B. Enderton. *A Mathematical Introduction to Logic*. Harcourt/Academic Press, San Diego, CA, 2nd edition, 2001.

[9] P. Gärdenfors. Conditionals and changes of belief. *Acta Philosophica Fennica*, 30(2–4):381–404, 1978.

[10] A. Gilio. Probabilistic reasoning under coherence in System P. *Annals of Mathematics and Artificial Intelligence*, 34(1):5–34, 2002.

[11] A. Gilio and G. Sanfilippo. Conditional random quantities and compounds of conditionals. *Studia Logica*, 102(4):709–729, 2014.

[12] J. Hawthorne and D. Makinson. The quantitative/qualitative watershed for rules of uncertain inference. *Studia Logica*, 86(2):247–297, 2007.

[13] K. Kelly and K. Genin. Learning, theory choice, and belief revision. *Submitted Manuscript*, 2016.

[14] S. Kraus, D. Lehmann, and M. Magidor. Nonmonotonic reasoning, preferential models and cumulative logics. *Artificial Intelligence*, 44(1–2):167–207, 1990.

[15] D. Lehmann and M. Magidor. What does a conditional knowledge base entail? *Artificial Intelligence*, 55(1):1–60, 1992.

[16] H. Leitgeb. *Inference on the Low Level*. Kluwer Academic Publishers, Dordrecht, Netherlands, 2004.

[17] H. Leitgeb. A probabilistic semantics for counterfactuals. Part A. *The Review of Symbolic Logic*, 5(1):26–84, 2012.

[18] H. Leitgeb. Reducing belief simpliciter to degrees of belief. *Annals of Pure and Applied Logic*, 164(12):1338–1389, 2013.

[19] H. Leitgeb. The stability theory of belief. *Philosophical Review*, 123(2):131–171, 2014.

[20] D. Lewis. *Counterfactuals*. Blackwell, Oxford, 1973.

[21] D. Lewis. Probabilities of conditionals and conditional probabilities. *The Philosophical Review*, 85(3):297–315, 1976.

[22] H. Lin and K. T. Kelly. A geo-logical solution to the lottery paradox, with applications to conditional logic. *Synthese*, 186(2):531–575, 2012.

[23] S. McCall. A history of connexivity. In D. Gabbay et al., editors, *A History of Logic*, volume 11, pages 415–449. Elsevier, Amsterdam, 2012.

[24] D. Nute. *Topics in Conditional Logic*. D. Reidel Publishing Company, Dordrecht, Netherlands, 1980.

[25] J. Pearl. *Probabilistic Reasoning in Intelligent Systems. Networks of Plausible Inference*. Morgan Kaufmann Publishers, San Francisco, 1988. Revised Second Printing.

[26] N. Pfeifer. Experiments on Aristotle's thesis. Towards an experimental philosophy of conditionals. *The Monist*, 95(2):223–240, 2012.

[27] N. Pfeifer and G. D. Kleiter. *The Conditional in Mental Probability Logic*, pages 153–173. Oxford University Press, Oxford, 2010.

[28] C. Pizzi and T. Williamson. Strong Boethius' Thesis and consequential implication. *Journal of Philosophical Logic*, 26(5):569–588, 1997.

[29] G. Priest. Negation as cancellation, and connexive logic. *Topoi*, 18(2):141–148, 1999.

[30] G Priest. *An Introduction to Non-Classical Logic*. Cambridge University Press, 2nd edition, 2008.

[31] H. Rott. Two concepts of plausibility in default reasoning. *Erkenntnis*, 79(6):1219–1252, 2014.

[32] G. Schurz. Probabilistic semantics for Delgrande's conditional logic and a counterexample to his default logic. *Artificial Intelligence*, 102(1):81–95, 1998.

[33] G. Schurz. Logic, matters of form, and closure under substitution. In L. Běhounek and M. Bílková, editors, *The Logica Yearbook 2004*, pages 33–46. Filosofia, Prague, 2004.

[34] G. Schurz and P. D. Thorn. Reward versus risk in uncertain inference. Theorems and simulations. *The Review of Symbolic Logic*, 5(4):574–612, 2012.

[35] K. Segerberg. Notes on conditional logic. *Studia Logica*, 48(2):157–168, 1989.

[36] Y. Shoham. A semantical approach to nonmonotonic logics. In M. Ginsberg, editor, *Readings in Nonmonotonic Reasoning*, pages 227–250. Morgan Kaufmann Publishers, Los Altos, CA, 1987.

[37] T. Skura. Refutation systems in propositional logic. In D. M. Gabbay and F. Guenthner, editors, *Handbook of Philosophical Logic*, volume 16, pages 115–157. Springer, Dordrecht, 2011.

[38] R. C. Stalnaker. A theory of conditionals. In N. Rescher, editor, *Studies in Logical Theory*, pages 98–112. Basil Blackwell, Oxford, 1968.

[39] M. Unterhuber. *Possible Worlds Semantics for Indicative and Counterfactual Conditionals? A Formal Philosophical Inquiry into Chellas-Segerberg Semantics*. Ontos, Frankfurt am Main, 2013.

[40] M. Unterhuber and G. Schurz. Completeness and correspondence in Chellas–Segerberg semantics. *Studia Logica*, 102(4):891–911, 2014.

[41] B.C. van Fraassen. Representational of conditional probabilities. *Journal of Philosophical Logic*, 5(3):417–430, 1976.

[42] H. Wansing. Connexive modal logic. In R. Schmidt et al., editors, *Advances in Modal Logic*, volume 5, pages 367–383. King's College Publications, London, 2005.

[43] H. Wansing. Connexive logic. In E. N. Zalta, editor, *The Stanford Encyclopedia of Philosophy*. Fall 2015 edition, 2015.

A Proof of Lemma 8

Lemma 8. AT and BT correspond to frame condition C_{AT} and C_{BT}, respectively.

Axiom schema AT. ('$\Rightarrow$'): Suppose a Chellas frame $\langle W, R \rangle$ does not satisfy C_{AT}. Thus, $\forall X \forall w \exists w'(wR_X w' \curlywedge w' \in X)$ does not hold and, consequently, there is a set of worlds $X \subseteq W$ and a world $w \in W$ such that either (i) there is no world $w' \in W$ such that $wR_X w'$ or (ii) there are only worlds $w' \in W$ such that $wR_X w'$

and $w' \notin X$. Choose V so that $\langle W, R, V \rangle$ is a Chellas model and $V(p) = X$. Then, either (i) there is no world $w' \in W$ such that $wR_{V(p)}w'$ or (ii) there are only worlds $w' \in W$ such that $wR_{V(p)}w'$ and $w' \notin V(p)$. If (i) is the case, trivially $\langle W, R, V \rangle \models_w p\,\square\!\!\rightarrow \neg p$ contradicting $\langle W, R \rangle \models_w \neg(p\,\square\!\!\rightarrow \neg p)$. Likewise, if (ii) holds, it follows that $\langle W, R, V \rangle \models_w p\,\square\!\!\rightarrow \neg p$, which contradicts $\langle W, R, V \rangle \models \neg(p\,\square\!\!\rightarrow \neg p)$ as well.

('$\Leftarrow$'): Let $\langle W, R \rangle$ be a Chellas frame such that $\langle W, R \rangle \not\models \neg(p\,\square\!\!\rightarrow \neg p)$. Then, there exists a formula α and a valuation function V such that $\langle W, R, V \rangle$ is a Chellas model as well as a world $w \in W$ such that $\langle W, R, V \rangle \models_w \alpha\,\square\!\!\rightarrow \neg\alpha$. It follows that for all worlds $w' \in W$ such that $wR_{V(\alpha)}w'$ it is the case that $\langle W, R, V \rangle \models_{w'} \neg\alpha$ where the latter is equivalent to $w' \notin V(\alpha)$. Thus, there is some $X \subseteq W$, namely $X = V(\alpha)$, and some $w \in W$ such that there is no world w' with $wR_X w'$ and $w' \in X$ contradicting $\forall X \forall w \exists w'(wR_X w' \wedge w' \in X)$.

$\square$

Axiom schema BT. ('$\Leftarrow$') Let $\langle W, R \rangle$ be a Chellas frame such that C_{BT} holds. Then, $\langle W, R \rangle$ satisfies $\forall X \forall w \exists w'(wR_X w')$. Suppose that $\langle W, R, V \rangle$ is a Chellas model such that $\langle W, R, V \rangle \models_w \alpha\,\square\!\!\rightarrow \beta$. Then, for all w'' such that $wR_{V(\alpha)}w''$ it is the case that $w'' \in V(\beta)$. As by C_{BT} for any $X \subseteq W$ there is a $w' \in W$ such that $wR_X w'$, we obtain $wR_{V(\alpha)}w'$ for every $V(\alpha) \subseteq W$. Consequently, as for all $w'' \in W$ such that $wR_{V(\alpha)}w''$ it is the case that $w'' \in V(\beta)$ – including w' – it follows that $w' \in V(\beta)$ and, thus, $w' \notin V(\neg\beta)$. Since $wR_{V(\alpha)}w'$ holds, this implies that $\langle W, R, V \rangle \not\models_w \alpha\,\square\!\!\rightarrow \neg\beta$ which gives us $\langle W, R, V \rangle \models (\alpha\,\square\!\!\rightarrow \beta) \rightarrow \neg(\alpha\,\square\!\!\rightarrow \neg\beta)$. Hence, the desired result is obtained.

('$\Rightarrow$') Let $\langle W, R \rangle$ be a Chellas frame such that C_{BT} does not hold. Then, there exists a world $w \in W$ and a set $X \subseteq W$ such that there is no world $w' \in W$ with $wR_X w'$. Let $\langle W, R, V \rangle$ be a Chellas model such that $X = V(p)$. Then, there exists no world $w' \in W$ such that $wR_{V(p)}w'$. It follows trivially that $\langle W, R, V \rangle \models_w p\,\square\!\!\rightarrow q$ and $\langle W, R, V \rangle \models_w p\,\square\!\!\rightarrow \neg q$. Consequently, $(p\,\square\!\!\rightarrow q) \rightarrow \neg(p\,\square\!\!\rightarrow \neg q)$ is not valid on $\langle W, R \rangle$.

$\square$

Natural Deduction for Bi-Connexive Logic and a Two-Sorted Typed λ-Calculus

Heinrich Wansing
Department of Philosophy II,
Ruhr-University Bochum, Germany
`heinrich.wansing@rub.de`

Abstract

The bi-connexive propositional logic **2C** is being introduced, which contains a connexive implication and a connexive co-implication that is in a certain sense dual to the connexive implication. The system **2C** is a connexive variant of the bi-intuitionistic logic **2Int** and contains a primitive strong negation. In both systems a relation of provability is supplemented with a certain relation of dual provability. Whereas entailment as the semantic counterpart of provability preserves support of truth from the premises to the conclusion of an inference, dual entailment as the semantic counterpart of dual provability preserves falsity from the premises to the conclusion of an inference. The strong negation that is added to the language of **2Int** to obtain the system **2C** internalizes falsification with respect to provability and it internalizes verification with respect to dual provability. This tight relation between verification, falsification, and strong negation allows one to see the dual of provability as disprovability.

After introducing a natural deduction proof system, N2C, and a relational semantics for **2C**, a two-sorted typed λ-calculus, 2λ, is presented that can be used to encode derivations in N2C. In particular, the {implication, co-implication, strong negation}–fragment of **2C** receives an encoding that makes use of functional application, functional abstraction, and certain sort/type-shift operations.

I would like to thank the audiences at the first workshop on connexive logics that was part of the *Fifth World Conference on Universal Logic* in Istanbul, June 2015, and the workshop *Proof theory of modal and nonclassical logics* during the *15th Congress on Logic, Methodology, and Philosophy of Science* (CLMPS) in Helsinki, August 2015, for encouraging feedback on an earlier version of this paper and two anonymous referees for their very constructive and helpful comments.

1 Introduction

In various branches of non-classical logic a distinction is drawn between truth and falsity as concepts that are primitive and independent of each other, though not necessarily disconnected. The separation of truth and falsity is achieved by giving up bivalence, so that falsity is distinguished from the absence of truth and truth is discriminated from the absence of falsity. In many-valued logic this leads to a distinction between a set of designated values and a set of antidesignated values and, moreover, to a multiplicity of entailment relations. In addition to the familiar preservation of designated values, there is, for example, q-entailment, "quasi-entailment", that leads from not-antidesignated premises to a designated conclusion (see [11, 12, 13]) and p-entailment, "plausibility-entailment", that leads from designated premises to a not-antidesignated conclusion (see [3, 4]). Quasi-entailment and plausibility-entailment are peculiar insofar as entailment is not defined in terms of preservation of membership in some subset of the set of truth values (alias truth degrees). In [23, p. 210], preservation of not being antidesignated from the premises to the conclusion of an inference is listed as an intuitively appealing notion of entailment. If truth (being designated) and falsity (being antidesignated) are treated on a par and not as each other's complement, then it makes much sense to take falsity-preservation from the premises to the conclusion of an inference very seriously as well.

In the logic of generalized truth values (see [22, 18], [23]), entailment relations are defined with respect to a partial order on a set of semantical values, in particular as relations defined on a set generated from the set **2** of classical truth values by iterated powerset formation. This approach very naturally leads to a distinction between a truth ordering and a separate falsity ordering on the powerset of the powerset of **2** together with two distinct entailment relations, truth entailment and falsity entailment.[1]

Whilst usually a notion of falsity is internalized into the logical object language by means of a negation connective, the notion of truth typically is *not* internalized at all. This is only one out of many ways in which "positive" concepts traditionally predominate even in non-classical logic in comparison to their "negative" counterparts. It would be possible to consider a formula A not as a vehicle for making an assertion, but rather as a device for making a denial. From the latter perspective one would be interested in having available a unary connective that internalizes truth instead of falsehood. The internalization of truth and falsity can be realized by a division of labour, namely by using two different negation connectives, or by

---

[1]Moreover, the subset relation on the set of generalized truth values may be seen as an information ordering.

utilizing a single one that internalizes both truth from the point of view of falsification and falsity from the perspective of verification. The former is achieved in the bi-intuitionistic logic **2Int** from [25]. In that system two negation operations are defined: intuitionistic negation and a connective that is in a certain sense dual to intuitionistic negation. Intuitionistic negation is negation as "implies falsity"; its dual, called co-negation, is understood as "co-implies truth". The intuitionistic negation $\neg A$ of a formula A internalizes an indirect notion of falsification into the logical object language from the point of view of verification: A state supports the truth of $\neg A$ iff A implies falsity (iff the assumption that A is true leads to the truth of the falsity constant $\bot$). The co-negation $-A$ of a formula A internalizes an indirect notion of verification into the object language from the point of view of falsification: A state supports the falsity of $-A$ iff A co-implies truth (iff the assumption that A is false leads to the falsity of the truth constant $\top$). The two negations are *defined* using the zero-place connectives $\top$ and $\bot$. One may, however, use a *primitive* strong negation, $\sim$, as in Nelson's constructive logics with strong negation **N3**, **N4** , and **N4**$^{\perp}$ (see, for example, [14, 15, 1, 27, 28, 2, 29, 31, 16, 17, 7, 8]) that provides both internalizations and thereby turns dual provability into a relation of disprovability, cf. [33].

Nevertheless, there is a certain preoccupation with the positive dimension of logic even in Nelson's systems and a lacuna in the separate treatment of truth and falsity. The constructive implication in Nelson's logics internalizes an entailment relation that preserves support of truth from the premises to the conclusion of an inference or, proof-theoretically, internalizes a corresponding derivability relation. However, in Nelson's logics there is no connective that internalizes the preservation of support of falsity from the premises to the conclusion of an inference or, proof-theoretically, internalizes a corresponding relation of dual derivability. Such a dual of implication, called co-implication, is present in the system **2Int** from [25, 36].[2] In the present paper, the bi-intuitionistic system **2Int** is modified. A primitive strong negation is added that internalizes falsity with respect to verification and truth with respect to falsification. Moreover, for this strong negation a connexive reading of negated

---

[2]An extension of Nelson's paraconsistent constructive logic **N4** by the co-implication connective of Rauszer's Heyting-Brouwer logic, called **HB** or **BiInt** (see [20, 6, 25] and references given there), is considered in [32, 33] and in [9]. This co-implication connective is dual to intuitionistic implication in another sense; it preserves non-truth from the conclusion of an inference to the conjunction of its premises. In the relational semantics of **BiInt**, a co-implication $A \prec B$ is said to be true at a state $w \in I$ from a relational model $\mathcal{M} = \langle I, \leq, v \rangle$ for **BiInt** iff there is a $u \in I$ with $u \leq w$, A is true at u, and B is not true at u: $\mathcal{M}, w \models A \prec B$ iff $(\exists u \in I)\, u \leq w$, $\mathcal{M}, u \models A$, and $\mathcal{M}, u \not\models B$. Moreover, $A \models B$ is defined to mean that for every model $\langle I, \leq, v \rangle$ and every $w \in I$, $\mathcal{M}, w \not\models B$ implies $\mathcal{M}, w \not\models A$. Let $\bot$ be a constantly untrue formula. Then $A \models B$ iff $A \prec B \models \bot$ because $\leq$ is reflexive.

implications and co-implications is assumed, so that the resulting bi-connexive logic **2C** also emerges as an extension of the connexive propositional logic **C** from [30], which was obtained from propositional **N4** by replacing the familiar falsification condition for strongly negated implications by its connexive version.[3]

Connexive logics have been motivated by quite different considerations, ranging from Aristotelian Syllogistic to Categorial Grammar and the semantics of indicative conditions, see, for example, [34]. Most prominently, the motivation is centered on the so-called Aristotle's Theses and Boethius' Theses, and the bi-connexive logic to be developed in the present paper allows on to prove these principles, see Section 2. A particular reason for considering *connexive* implication, $\rightarrow$, and *connexive* co-implication, $\prec$, instead of assuming the familiar understanding of negated implications in Nelson's and other logics is that one obtains a neat encoding of derivations in the $\{\rightarrow, \prec, \sim\}$-fragment of the language under consideration by typed λ-terms built up from atomic terms of two sorts, one for proofs and one for dual proofs, using only (i) functional application, (ii) functional abstraction, and (iii) certain sort/type-shift operations that turn an encoding of a dual proof of a formula A [respectively $\sim A$] into an encoding of a proof of $\sim A$ [respectively A] and that turn an encoding of a proof of a formula A [respectively $\sim A$] into an encoding of a dual proof of $\sim A$ [respectively A]. In [26, 28] an encoding of derivations in Nelson's constructive logics with strong negation **N3** and **N4** was obtained by giving up the unique typedness of terms. The use of terms of two sorts avoids this feature: Every term is uniquely typed.

2 The natural deduction proof system N2C

Syntactically, the language $\mathcal{L}_{2C}$ of the bi-connexive system **2C** extends the language of the bi-intuitionistic logic **2Int** by a primitive negation connective $\sim$ and is defined in Backus–Naur form as follows:

$$A ::= \ p \mid \bot \mid \top \mid \sim A \mid (A \wedge A) \mid (A \vee A) \mid (A \rightarrow A) \mid (A \prec A).$$

---

[3]An anonymous referee expounded that even before a connexive reading of implications is discussed, the notions of co-implication and dual proof need to be conceptually clarified. Here I would just like to emphasize that if falsity is seen as an independent notion in its own right and is taken to be on a par with truth, then support–of–falsity preservation arises quite naturally as a companion to the preservation of support of truth. The object-language counterpart of support–of–falsity preservation (from the premises to the conclusion of a dually valid inference) is co-implication as understood in **2Int** and **2C**. Conceptual clarification in addition to the motivating considerations in this introductory section can be found, for example, in [32, 33, 25, 35]. A connection between dual intuitionistic logic and Popperian philosophy of science is discussed in [21].

where p is a propositional variable from some fixed infinite set Φ of sentential variables (atomic formulas). The language $\mathcal{L}_{\mathbf{Int}}$ of intuitionistic propositional logic, **Int** (cf. [24]), is $\mathcal{L}_{\mathbf{2C}}$ restricted to the connectives $\top$, $\bot$, $\wedge$, $\vee$, and $\rightarrow$; the language $\mathcal{L}_{\mathbf{DualInt}}$ of dual intuitionistic propositional logic, **DualInt** (see, [6, 21, 25]), is $\mathcal{L}_{\mathbf{2C}}$ restricted to the connectives $\top$, $\bot$, $\wedge$, $\vee$, and $\prec$. We write $A \equiv B$ to indicate that the formulas A and B are identical symbols.

In **2C** (and in **2Int**), the co-implication connective $\prec$ is in a sense dual to intuitionistic implication, it internalizes a relation of dual derivability into the logical object language. Dual derivability leads from counterassumptions (premises assumed to be false) to false conclusions. The language of **2C** (and that of **2Int**) allows one to define two "weak" negation connectives; the co-negation $-A$ of A is defined as $\top \prec A$, and the more familiar intuitionistic negation $\neg A$ of A is defined as $A \rightarrow \bot$. Moreover, the equivalence and co-equivalence connectives $\leftrightarrow$ and $\succ\!\!\prec$ are defined as follows:

$$(A \leftrightarrow B) := ((A \rightarrow B) \wedge (B \rightarrow A)); \quad (A \succ\!\!\prec B) := ((A \prec B) \vee (B \prec A)).$$

A sound and complete natural deduction proof system N2Int for **2Int** is presented in [25, 36]. The system uses single-line rules for proofs and double-line rules for dual proofs. Derivations in N2Int combine proofs and dual proofs, so that a proof, in which the conclusion appears under a single line, may contain dual proofs as subderivations, and a dual proof, in which the conclusion appears under a double line, may contain proofs as subderivations. The conclusions of proofs and dual proofs depend on ordered pairs $(\Delta; \Gamma)$ of finite sets of premises, a set Δ of assumptions that are taken to be true, and a set Γ of "counterassumptions" that are taken to be false. Single square brackets $[\,]$ are used to indicate that assumptions may be cancelled, and double-square brackets $[\![\,]\!]$ are used to indicate that counterassumptions may be cancelled.[4] The proof system N2C we are about to define shares these features with N2Int.

The proof rules for the connectives $\top$, $\bot$, $\wedge$, $\vee$, and $\rightarrow$ are more or less those of intuitionistic logic, and the rules for introducing or eliminating $\top$, $\bot$, $\wedge$, and $\vee$ into or from dual proofs are obtained by a dualization of their introduction and elimination rules for proofs. The dual proof rules for implication are such that together with the proof and dual proof rules for strong negation, implication is a connexive implication, so that the characteristic theorems of systems of connexive logic (cf. [34] and the references therein), namely the so-called theses of Aristotle and Boethius, are provable. The introduction and elimination rules for $\prec$ are obtained

---

[4]We usually write $[A]$ instead of $[\overline{A}]$ and $[\![A]\!]$ instead of $[\![\overline{\overline{A}}]\!]$.

by dualizing the rules for $\rightarrow$, so that duals of Aristotle's theses and Boethius' theses are dually provable (see p. 6).

Definition 1. We consider $\overline{A}$ as a proof of A from $(\{A\}; \varnothing)$ and $\overline{\overline{A}}$ as a dual proof of A from $(\varnothing; \{A\})$. Moreover $\overline{\top}$ is a proof of $\top$ from $(\varnothing; \varnothing)$ and $\overline{\overline{\bot}}$ is a dual proof of $\bot$ from $(\varnothing; \varnothing)$. In addition to these stipulations, the system N2C comprises the introduction and elimination rules from Tables 1 and 2, where Ep stands for "elimination from proofs", Ip for "introduction into proofs", Edp for "elimination from dual proofs", and Idp for "introduction into dual proofs".

We write $(\Delta; \Gamma) \vdash A$ if there is a proof of A from $(\Delta; \Gamma)$; and we write $(\Delta; \Gamma) \vdash^d A$ if there is a dual proof of A from $(\Delta; \Gamma)$. Moreover, we assume that if $(\Delta; \Gamma) \vdash A$, $\Delta \subseteq \Delta'$ and $\Gamma \subseteq \Gamma'$ for finite sets of $\mathcal{L}_{2\mathrm{Int}}$-formulas Δ' and Γ', then $(\Delta'; \Gamma') \vdash A$. Similarly, we assume that if $(\Delta; \Gamma) \vdash^d A$, $\Delta \subseteq \Delta'$ and $\Gamma \subseteq \Gamma'$ for finite sets of $\mathcal{L}_{2\mathrm{C}}$-formulas Δ' and Γ', then $(\Delta'; \Gamma') \vdash^d A$. We write $\mathsf{P}(\Pi, A, (\Delta; \Gamma))$ $[\mathsf{DP}(\Pi, A, (\Delta; \Gamma))]$ if Π is a proof [dual proof] of A from the set of assumptions Δ and the set of counterassumptions Γ. The predicates P and DP can be used in an inductive definition of proofs and dual proofs. The rule $(\wedge Edp)$, for example, is then captured by the following clause: If $\mathsf{DP}(\Pi, (A \wedge B), (\Delta; \Gamma))$, $\mathsf{DP}(\Pi', C, (\Delta'; \Gamma' \cup \{A\}))$ and $\mathsf{DP}(\Pi'', C, (\Delta''; \Gamma'' \cup \{B\}))$, then

$$\mathsf{DP}(\overset{\Pi \ \Pi' \ \Pi''}{\overline{C}}, C, (\Delta \cup \Delta' \cup \Delta''; \Gamma \cup (\Gamma' \setminus \{A\}) \cup (\Gamma'' \setminus \{B\}))).$$

This presentation states the sets of assumptions and counterassumptions of the dual proof obtained by applying $(\wedge Edp)$ explicitly.

In the proof system N2C, Aristotle's theses $\sim(A \rightarrow \sim A)$, $\sim(\sim A \rightarrow A)$, Boethius' theses, $\sim(A \rightarrow B) \rightarrow (A \rightarrow \sim B)$, $(A \rightarrow \sim B) \rightarrow \sim(A \rightarrow B)$, and their duals $\sim(\sim A \prec A)$, $\sim(A \prec \sim A)$, $(\sim B \prec A) \prec \sim(B \prec A)$, $\sim(B \prec A) \prec (\sim B \prec A)$ are provable, respectively dually provable as follows:

$$
\frac{\dfrac{[A]}{\overline{\overline{\sim A}}}}{\dfrac{A \rightarrow \sim A}{\sim(A \rightarrow \sim A)}}
\qquad
\frac{\dfrac{[\sim A]}{\overline{\overline{A}}}}{\dfrac{\sim A \rightarrow A}{\sim(\sim A \rightarrow A)}}
\qquad
\frac{\dfrac{[\![A]\!]}{\overline{\overline{\sim A}}}}{\dfrac{\sim A \prec A}{\sim(\sim A \prec A)}}
\qquad
\frac{\dfrac{[\![A]\!]}{\overline{\overline{A}}}}{\dfrac{A \prec \sim A}{\sim(A \prec \sim A)}}
$$

$$
\frac{\dfrac{[A] \quad \dfrac{[\sim(A \rightarrow B)]}{A \rightarrow B}}{\dfrac{B}{\dfrac{\sim B}{A \rightarrow \sim B}}}}{\sim(A \rightarrow B) \rightarrow (A \rightarrow \sim B)}
\qquad
\frac{\dfrac{[A] \quad [A \rightarrow \sim B]}{\dfrac{\sim B}{\dfrac{B}{\dfrac{A \rightarrow B}{\sim(A \rightarrow B)}}}}}{(A \rightarrow \sim B) \rightarrow \sim(A \rightarrow B)}
$$

$$\frac{\begin{array}{c}(\Delta;\Gamma)\\ \vdots\\ \bot\end{array}}{A}\ (\bot Ep) \qquad \frac{\begin{array}{c}(\Delta;\Gamma)\\ \vdots\\ A\end{array}}{{\sim}A}\ ({\sim}Ip) \qquad \frac{\begin{array}{c}(\Delta;\Gamma)\\ \vdots\\ {\sim}A\end{array}}{A}\ ({\sim}Ep)$$

$$\frac{\begin{array}{cc}(\Delta;\Gamma) & (\Delta';\Gamma')\\ \vdots & \vdots\\ A & B\end{array}}{(A\wedge B)}\ (\wedge Ip) \qquad \frac{\begin{array}{c}(\Delta;\Gamma)\\ \vdots\\ (A\wedge B)\end{array}}{A}\ (\wedge Ep) \qquad \frac{\begin{array}{c}(\Delta;\Gamma)\\ \vdots\\ (A\wedge B)\end{array}}{B}\ (\wedge Ep)$$

$$\frac{\begin{array}{c}(\Delta;\Gamma)\\ \vdots\\ A\end{array}}{(A\vee B)}\ (\vee Ip) \qquad \frac{\begin{array}{c}(\Delta;\Gamma)\\ \vdots\\ B\end{array}}{(A\vee B)}\ (\vee Ip)$$

$$\frac{\begin{array}{ccc}(\Delta;\Gamma) & ([A],\Delta';\Gamma') & ([B],\Delta'';\Gamma'')\\ \vdots & \vdots & \vdots\\ (A\vee B) & C & C\end{array}}{C}\ (\vee Ep)$$

$$\frac{\begin{array}{c}([A],\Delta;\Gamma)\\ \vdots\\ B\end{array}}{(A\to B)}\ (\to Ip) \qquad \frac{\begin{array}{cc}(\Delta;\Gamma) & (\Delta';\Gamma')\\ \vdots & \vdots\\ A & (A\to B)\end{array}}{B}\ (\to Ep)$$

$$\frac{\begin{array}{c}(\Delta;\Gamma,[\![A]\!])\\ \vdots\\ B\end{array}}{(B{\prec}A)}\ ({\prec}Ip) \qquad \frac{\begin{array}{cc}(\Delta';\Gamma') & (\Delta;\Gamma)\\ \vdots & \vdots\\ (B{\prec}A) & A\end{array}}{B}\ ({\prec}Ep)$$

Table 1: Introduction and elimination rules of N2C w.r.t. proofs.

$$\frac{(\Delta;\Gamma)}{\overline{\overline{\top}}} \ (\top Edp)$$

$$\frac{(\Delta;\Gamma)}{\overline{A}} \ (\sim Idp)$$

$$\frac{(\Delta;\Gamma)}{\overline{\sim A}} \ (\sim Edp)$$

$$\frac{(\Delta;\Gamma) \quad (\Delta';\Gamma')}{\overline{\overline{A}} \quad \overline{\overline{B}}} \ (\vee Idp)$$

$$\frac{(\Delta;\Gamma)}{\overline{\overline{(A \vee B)}}} \ (\vee Edp)$$

$$\frac{(\Delta;\Gamma)}{\overline{\overline{(A \vee B)}}} \ (\vee Edp)$$

$$\frac{(\Delta;\Gamma)}{\overline{\overline{A}}} \ (\wedge Idp)$$

$$\frac{(\Delta;\Gamma)}{\overline{\overline{B}}} \ (\wedge Idp)$$

$$\frac{(\Delta;\Gamma) \quad (\Delta';\Gamma',[\![A]\!]) \quad (\Delta'';\Gamma'',[\![B]\!])}{\overline{\overline{(A \wedge B)}} \quad \overline{\overline{C}} \quad \overline{\overline{C}}} \ (\wedge Edp)$$

$$\frac{(\Delta;\Gamma,[\![A]\!])}{\overline{\overline{B}}} \ (\prec Idp)$$

$$\frac{(\Delta';\Gamma') \quad (\Delta;\Gamma)}{\overline{\overline{(B \prec A)}} \quad \overline{\overline{A}}} \ (\prec Edp)$$

$$\frac{([A],\Delta;\Gamma)}{\overline{\overline{B}}} \ (\to Idp)$$

$$\frac{(\Delta;\Gamma) \quad (\Delta';\Gamma')}{\overline{A} \quad \overline{\overline{(A \to B)}}} \ (\to Edp)$$

Table 2: Introduction and elimination rules of N2C w.r.t. dual proofs.

420

$$\frac{\dfrac{\dfrac{\dfrac{\dfrac{[\![\sim(B\!-\!\prec\!A)]\!]}{B\!-\!\prec\!A}\quad[\![A]\!]}{B}}{\sim\!B}}{\sim\!B\!-\!\prec\!A}}{(\sim\!B\!-\!\prec\!A)\!-\!\prec\!\sim\!(B\!-\!\prec\!A)} \qquad \frac{\dfrac{\dfrac{\dfrac{\dfrac{[\sim\!B\!-\!\prec\!A]\quad[\![A]\!]}{\sim\!B}}{B}}{B\!-\!\prec\!A}}{\sim\!(B\!-\!\prec\!A)}}{\sim\!(B\!-\!\prec\!A)\!-\!\prec\!(\sim\!B\!-\!\prec\!A)}$$

The strong negation in **2C** operates as a switch between provability and disprovability, and the distinction between assumptions and counterassumptions could be dispensed with in **2C**. Let $\sim\!\Theta := \{\sim\!A \mid A \in \Theta\}$ for a set of formulas Θ. If $\Theta = \varnothing$, let $\sim\!\Theta := \varnothing$. Then

(*) $(\Delta;\Gamma) \vdash A$ iff $(\Delta \cup \sim\!\Gamma; \varnothing) \vdash A$, $(\Delta;\Gamma) \vdash^d A$ iff $(\varnothing; \Gamma \cup \sim\!\Delta) \vdash^d A$.

(**) $(\Delta;\Gamma) \vdash A$ iff $(\Delta;\Gamma) \vdash^d \sim\!A$, $(\Delta;\Gamma) \vdash^d A$ iff $(\Gamma;\Delta) \vdash \sim\!A$.

Still, if we consider $\{\sim, \bot\}$-free or $\{\sim, \top\}$-free fragments of $\mathcal{L}_{\mathbf{2C}}$, the distinction between assumptions and counterassumptions and between provability and its dual is indispensable.

Let $A(B)$ stand for the result of the uniform replacement of all occurrences of a certain propositional variable in A by B. It is well-known that in Nelson's constructive logics **N3**, **N4**, and **N4**$^\perp$ with strong negation, provable equivalence is not a congruence relation, i.e., these systems are not closed under the replacement rule:

$$\frac{\varnothing \vdash A \leftrightarrow B}{\varnothing \vdash C(A) \leftrightarrow C(B).}$$

Often this is shown by pointing out that in Nelson's logics the usual understanding of negated implications is assumed, so that $\sim(p \to q) \leftrightarrow (p \wedge \sim\!q)$ is provable, whereas $\sim\!\sim(p \to q) \leftrightarrow \sim(p \wedge \sim\!q)$ fails to be provable, for otherwise $(p \to q) \leftrightarrow (\sim\!p \vee q)$ would be provable. Nelson's logics are closed under the restriction of the replacement rule to $\sim$-free formulas C and under the weak replacement rule

$$\frac{\varnothing \vdash A \Leftrightarrow B}{\varnothing \vdash C(A) \Leftrightarrow C(B),}$$

where the strong equivalence $A \Leftrightarrow B$ is defined as $(A \leftrightarrow B) \wedge (\sim\!A \leftrightarrow \sim\!B)$. Since we assume a connexive understanding of strongly negated implications, the above counterexample does not work. In the paraconsistent logics **N4** and **N4**$^\perp$ there are, however, other counterexamples. For atomic formulas p and q, the equivalence $(p \to p) \leftrightarrow (q \to q)$ is provable, whereas $\sim(p \to p) \leftrightarrow \sim(q \to q)$ is not, which gives us a counterexample for **2C** as well. Let strong co-equivalence $A \succ\!\!=\!\!\prec B$ be defined as $(A \succ\!\!\prec B) \vee (\sim\!A \succ\!\!\prec \sim\!B)$.

421

Proposition 1. The system N2C is closed under the following replacement rules:

$$\frac{(\varnothing;\varnothing) \vdash A \Leftrightarrow B}{(\varnothing;\varnothing) \vdash C(A) \Leftrightarrow C(B)} \qquad \frac{(\varnothing;\varnothing) \vdash^d A \rightarrowtail\leftarrow B}{(\varnothing;\varnothing) \vdash^d C(A) \rightarrowtail\leftarrow C(B)}$$

Proof. By induction on C. The cases where C is one of the constants $\top$, $\bot$ or an atomic formula are trivial. If C is a strong negation $\sim D$, we use induction on D; the case where D is atomic and thus $D_A \equiv \sim A$ and $D_B \equiv \sim B$ is almost trivial. We have, for example:

$$\frac{\dfrac{\dfrac{[\sim\sim A]}{\sim A}}{\dfrac{A}{}}\quad \dfrac{\dfrac{(\varnothing;\varnothing)}{\vdots}{A \Leftrightarrow B}}{\vdots}}{\dfrac{\dfrac{B}{\sim B}}{\dfrac{\sim\sim B}{\sim\sim A \to \sim\sim B}}}$$

$$\frac{\dfrac{\dfrac{(\varnothing;\varnothing)}{\vdots}{A \rightarrowtail\leftarrow B}}{\dfrac{\vdots}{B \rightarrow\!\!\!\prec A}}\quad \dfrac{[\![\sim\sim A]\!]}{\dfrac{\sim A}{A}}}{\dfrac{\dfrac{B}{\sim B}}{\dfrac{\sim\sim B}{\sim\sim B \rightarrow\!\!\!\prec \sim\sim A}}}$$

The other cases are also straightforward. We here present some derivations that are relevant to case $C \equiv (E \rightarrow\!\!\!\prec D)$:

$$\frac{\dfrac{\dfrac{(\varnothing;\varnothing)}{\vdots}{E_A \Leftrightarrow E_B}}{\dfrac{\vdots}{\sim E_A \to \sim E_B}}\quad \dfrac{\dfrac{[\sim(E_A \rightarrow\!\!\!\prec D_A)]}{E_A \rightarrow\!\!\!\prec D_A}}{\dfrac{\dfrac{\dfrac{(\varnothing;\varnothing)}{\vdots}{D_A \Leftrightarrow D_B}}{\dfrac{\vdots}{\sim D_B \to \sim D_A}}\quad \dfrac{[\![D_B]\!]}{\sim D_B}}{\dfrac{\sim D_A}{D_A}}}}{\dfrac{\dfrac{E_A}{\sim E_A}}{\dfrac{\dfrac{\sim E_B}{E_B}}{\dfrac{E_B \rightarrow\!\!\!\prec D_B}{\dfrac{\sim(E_B \rightarrow\!\!\!\prec D_B)}{\sim(E_A \rightarrow\!\!\!\prec D_A) \to \sim(E_B \rightarrow\!\!\!\prec D_B)}}}}}$$

$$
\cfrac{
\cfrac{(\varnothing;\varnothing)}{\vdots} \qquad
\cfrac{
\cfrac{[\![\sim(E_B \prec D_B)]\!]}{E_B \prec D_B} \qquad
\cfrac{
\cfrac{(\varnothing;\varnothing)}{\vdots} \\ \cfrac{D_A \Leftrightarrow D_B}{\vdots} \\ \cfrac{\sim D_A \to \sim D_B \quad \cfrac{[\![D_A]\!]}{\sim D_A}}{\cfrac{\sim D_B}{D_B}}
}{E_B}
}{
\cfrac{E_A}{\cfrac{E_A \prec D_A}{\sim(E_A \prec D_A)}}
}
}{\sim(E_A \prec D_A) \prec \sim(E_B \prec D_B)}
$$

□

Proposition 2. In **2C**, (i) $\wedge$ is definable in terms of $\vee$ and $\sim$, (ii) $\vee$ is definable in terms of $\wedge$ and $\sim$, (iii) $\to$ is definable in terms of $\prec$ and $\sim$, and (iv) $\prec$ is definable in terms of $\to$ and $\sim$.

Proof. In view of Proposition 1, it is enough to note the following facts:

$$(\varnothing;\varnothing) \vdash (A \wedge B) \leftrightarrow \sim(\sim A \vee \sim B), \quad (\varnothing;\varnothing) \vdash^d (A \wedge B) \succ\!\!\prec \sim(\sim A \vee \sim B),$$

$$(\varnothing;\varnothing) \vdash (A \vee B) \leftrightarrow \sim(\sim A \wedge \sim B), \quad (\varnothing;\varnothing) \vdash^d (A \vee B) \succ\!\!\prec \sim(\sim A \wedge \sim B),$$

$$(\varnothing;\varnothing) \vdash (A \to B) \leftrightarrow (B \prec \sim A), \quad (\varnothing;\varnothing) \vdash^d (A \to B) \succ\!\!\prec (B \prec \sim A),$$

$$(\varnothing;\varnothing) \vdash (B \prec A) \leftrightarrow (\sim A \to B), \quad (\varnothing;\varnothing) \vdash^d (B \prec A) \succ\!\!\prec (\sim A \to B)$$

$$(\varnothing;\varnothing) \vdash \sim(A \wedge B) \leftrightarrow \sim\sim(\sim A \vee \sim B), \quad (\varnothing;\varnothing) \vdash^d \sim(A \wedge B) \succ\!\!\prec \sim\sim(\sim A \vee \sim B),$$

$$(\varnothing;\varnothing) \vdash \sim(A \vee B) \leftrightarrow \sim\sim(\sim A \wedge \sim B), \quad (\varnothing;\varnothing) \vdash^d \sim(A \vee B) \succ\!\!\prec \sim\sim(\sim A \wedge \sim B),$$

$$(\varnothing;\varnothing) \vdash \sim(A \to B) \leftrightarrow \sim(B \prec \sim A), \quad (\varnothing;\varnothing) \vdash^d \sim(A \to B) \succ\!\!\prec \sim(B \prec \sim A),$$

$$(\varnothing;\varnothing) \vdash \sim(B \prec A) \leftrightarrow \sim(\sim A \to B), \quad (\varnothing;\varnothing) \vdash^d \sim(B \prec A) \succ\!\!\prec \sim(\sim A \to B).$$

□

3   Models for 2C

The relational model theory of **2C** is a modification of the relational model theory of **2Int** in [25].

Definition 2. A model for **2C** is a structure $\mathcal{M} = \langle I, \leq, v^+, v^- \rangle$, where I is a non-empty set of states, $\langle I, \leq \rangle$ is a pre-order, and v^+, v^- are valuation functions from

Φ to subsets of I. Instead of $x \leq x'$ we also write $x' \geq x$. For $x \in I$ the relations $\mathcal{M}, x \models^+ A$ ("x supports the truth of A in $\mathcal{M}$") and $\mathcal{M}, x \models^- A$ ("x supports the falsity of A in $\mathcal{M}$") are inductively defined as follows:

$$\mathcal{M}, x \models^+ \top \quad \mathcal{M}, x \not\models^- \top \quad \mathcal{M}, x \not\models^+ \bot \quad \mathcal{M}, x \models^- \bot$$

$$
\begin{aligned}
\mathcal{M}, x &\models^+ p & &\text{iff} & x &\in v^+(p) \\
\mathcal{M}, x &\models^- p & &\text{iff} & x &\in v^-(p) \\[4pt]
\mathcal{M}, x &\models^+ {\sim}A & &\text{iff} & \mathcal{M}, x &\models^- A \\
\mathcal{M}, x &\models^- {\sim}A & &\text{iff} & \mathcal{M}, x &\models^+ A \\[4pt]
\mathcal{M}, x &\models^+ (A \wedge B) & &\text{iff} & \mathcal{M}, x &\models^+ A \text{ and } \mathcal{M}, x \models^+ B \\
\mathcal{M}, x &\models^- (A \wedge B) & &\text{iff} & \mathcal{M}, x &\models^- A \text{ or } \mathcal{M}, x \models^- B \\[4pt]
\mathcal{M}, x &\models^+ (A \vee B) & &\text{iff} & \mathcal{M}, x &\models^+ A \text{ or } \mathcal{M}, x \models^+ B \\
\mathcal{M}, x &\models^- (A \vee B) & &\text{iff} & \mathcal{M}, x &\models^- A \text{ and } \mathcal{M}, x \models^- B \\[4pt]
\mathcal{M}, x &\models^+ (A \to B) & &\text{iff} & &\text{for every } x' \geq x : \mathcal{M}, x' \not\models^+ A \text{ or } \mathcal{M}, x' \models^+ B \\
\mathcal{M}, x &\models^- (A \to B) & &\text{iff} & &\text{for every } x' \geq x : \mathcal{M}, x' \not\models^+ A \text{ or } \mathcal{M}, x' \models^- B \\[4pt]
\mathcal{M}, x &\models^+ (B {\prec} A) & &\text{iff} & &\text{for every } x' \geq x : \mathcal{M}, x' \not\models^- A \text{ or } \mathcal{M}, x' \models^+ B \\
\mathcal{M}, x &\models^- (B {\prec} A) & &\text{iff} & &\text{for every } x' \geq x : \mathcal{M}, x' \not\models^- A \text{ or } \mathcal{M}, x' \models^- B.
\end{aligned}
$$

Moreover, support of truth and support of falsity are required to be persistent for atomic formulas. For every $p \in \Phi$, and all states x, x': if $x \leq x'$ and $\mathcal{M}, x \models^+ p$, then $\mathcal{M}, x' \models^+ p$ and if $x \leq x'$ and $\mathcal{M}, x \models^- p$, then $\mathcal{M}, x' \models^- p$.

Proposition 3. For every $\mathcal{L}_{2C}$-formula A, and all states x, x' of any model $\mathcal{M}$ for **2C** $\mathcal{M}$: if $x \leq x'$ and $\mathcal{M}, x \models^+ A$, then $\mathcal{M}, x' \models^+ A$ and if $x \leq x'$ and $\mathcal{M}, x \models^- A$, then $\mathcal{M}, x' \models^- A$.

Proof. By induction on A. $\qquad\qquad\qquad\qquad\qquad\qquad\qquad\qquad\qquad\qquad$ $\square$

Definition 3. An $\mathcal{L}_{2C}$-formula A is valid in a model for **2C** $\mathcal{M} = \langle I, \leq, v^+, v^- \rangle$ iff for every $x \in I$, $\mathcal{M}, x \models^+ A$; A is valid in **2C** ($\models_{2C} A$) iff A is valid in every model for **2C**.

An $\mathcal{L}_{2C}$-formula A is dually valid in a model for **2C** $\mathcal{M} = \langle I, \leq, v^+, v^- \rangle$ iff for every $x \in I$, $\mathcal{M}, x \models^- A$; A is dually valid in **2C** ($\models^d_{2C} A$) iff A is dually valid in every model for **2C**.

Definition 4. Let $\Delta \cup \{A\}$ be a set of $\mathcal{L}_{2C}$-formulas. Δ entails A ($\Delta \models A$) iff for every model for **2C** $\mathcal{M} = \langle I, \leq, v^+, v^- \rangle$ and every $x \in I$, it holds that if the truth of every element of Δ is supported by x, then the truth of A is supported by x.

Let $\Delta \cup \{A\}$ be a set of $\mathcal{L}_{\mathbf{2C}}$-formulas. Δ dually entails A ($\Delta \models^d A$) iff for every model for $\mathbf{2C}$ $\mathcal{M} = \langle I, \leq, v^+, v^- \rangle$ and every $x \in I$, it holds that if the falsity of every element of Δ is supported by x, then the falsity of A is supported by x.

Definition 5. If $\langle I, \leq, v^+, v^- \rangle$ is a model for $\mathbf{2C}$, then $\mathcal{M} = \langle I, \leq, v^+ \rangle$ is a model for **Int**. Formulas from $\mathcal{L}_{\mathbf{Int}}$ are interpreted in models for **Int** as in models for $\mathbf{2C}$, i.e., for an $\mathcal{L}_{\mathbf{Int}}$-formula A the relation $\mathcal{M}, x \models^+ A$ ("x supports the truth of A in $\mathcal{M}$") is defined as in Definition 2. Moreover, support of truth is required to be persistent for atomic formulas.

An $\mathcal{L}_{\mathbf{Int}}$-formula A is valid in a model for **Int** $\mathcal{M} = \langle I, \leq, v^+ \rangle$ iff for every $w \in I$, $\mathcal{M}, w \models^+ A$; A is valid in **Int** ($\models_{\mathbf{Int}} A$) iff it is valid in every model for **Int**.

Definition 6. If $\langle I, \leq, v^+, v^- \rangle$ is a model for $\mathbf{2C}$, then $\mathcal{M} = \langle I, \leq, v^- \rangle$ is a model for **DualInt**. Formulas from $\mathcal{L}_{\mathbf{DualInt}}$ are interpreted in models for **DualInt** as in models for $\mathbf{2C}$, i.e., for an $\mathcal{L}_{\mathbf{DualInt}}$-formula A the relation $\mathcal{M}, x \models^- A$ ("x supports the falsity of A in $\mathcal{M}$") is defined as in Definition 2. Moreover, support of falsity is required to be persistent for atomic formulas.

An $\mathcal{L}_{\mathbf{DualInt}}$-formula A is dually valid in a model $\mathcal{M} = \langle I, \leq, v^- \rangle$ for **DualInt** iff for every $w \in I$, $\mathcal{M}, w \models^- A$; A is dually valid in **DualInt** ($\models^d_{\mathbf{DualInt}} A$) iff it is dually valid in every model for **DualInt**.

In [25] two recursive translation functions τ and ζ are defined. The function τ translates formulas of $\mathbf{2Int}$ into formulas of **Int**, and ζ maps formulas of $\mathbf{2Int}$ to formulas of **DualInt**. It is then shown that τ is a faithful embedding of $\mathbf{2Int}$ into **Int** with respect to entailment, and that ζ is a faithful embedding of $\mathbf{2Int}$ into **DualInt** with respect to dual entailment. In the case of $\mathbf{2C}$, we define similar translations, here again denoted by τ and ζ.

Definition 7. Let $\Phi' = \{p' \mid p \in \Phi\}$. We inductively define the translations τ from $\mathcal{L}_{\mathbf{2C}}$ into $\mathcal{L}_{\mathbf{Int}}$ and ζ from $\mathcal{L}_{\mathbf{2C}}$ into $\mathcal{L}_{\mathbf{DualInt}}$ based on the set of atomic formulas $\Phi \cup \Phi'$ as follows (some outermost brackets are omitted):

$$
\begin{aligned}
\tau(p) &:= p & \tau(\sim p) &:= p' \\
\tau(\top) &:= \top & \tau(\sim\top) &:= \bot \\
\tau(\bot) &:= \bot & \tau(\sim\bot) &:= \top \\
\tau(A \wedge B) &:= \tau(A) \wedge \tau(B) & \tau(\sim(A \wedge B)) &:= \tau(\sim A) \vee \tau(\sim B) \\
\tau(A \vee B) &:= \tau(A) \vee \tau(B) & \tau(\sim(A \vee B)) &:= \tau(\sim A) \wedge \tau(\sim B) \\
\tau(A \to B) &:= \tau(A) \to \tau(B) & \tau(\sim(A \to B)) &:= \tau(A) \to \tau(\sim B) \\
\tau(B \prec A) &:= \tau(\sim A) \to \tau(B) & \tau(\sim(B \prec A)) &:= \tau(\sim A) \to \tau(\sim B) \\
& & \tau(\sim\sim A) &:= \tau(A)
\end{aligned}
$$

$$\begin{aligned}
\zeta(p) &:= p & \zeta(\sim p) &:= p' \\
\zeta(\top) &:= \top & \zeta(\sim\top) &:= \bot \\
\zeta(\bot) &:= \bot & \zeta(\sim\bot) &:= \top \\
\zeta(A \wedge B) &:= \zeta(A) \wedge \zeta(B) & \zeta(\sim(A \wedge B)) &:= \zeta(\sim A) \vee \zeta(\sim B) \\
\zeta(A \vee B) &:= \zeta(A) \vee \zeta(B) & \zeta(\sim(A \vee B)) &:= \zeta(\sim A) \wedge \zeta(\sim B) \\
\zeta(A \to B) &:= \zeta(B) \prec \zeta(\sim A) & \zeta(\sim(A \to B)) &:= \zeta(\sim B) \prec \zeta(\sim A) \\
\zeta(B \prec A) &:= \zeta(B) \prec \zeta(A) & \zeta(\sim(B \prec A)) &:= \zeta(\sim B) \prec \zeta(A) \\
& & \zeta(\sim\sim A) &:= \zeta(A)
\end{aligned}$$

An $\mathcal{L}_{2C}$-formula A is in negation normal form iff it contains occurrences of $\sim$ only as prefixes of atomic formulas.

Proposition 4. *For every $\mathcal{L}_{2C}$-formula A, there is a formula A' in negation normal form such that (i) $\models_{2C} A \leftrightarrow A'$ and (ii) $\models_{2C}^d A \succ\!\!\prec A'$.*

Proof. The following equivalences are *valid* in **2C**:

$$\sim\top \leftrightarrow \bot, \quad \sim\bot \leftrightarrow \top, \qquad \sim\sim A \leftrightarrow A$$
$$\sim(A \wedge B) \leftrightarrow (\sim A \vee \sim B), \quad \sim(A \vee B) \leftrightarrow (\sim A \wedge \sim B)$$
$$\sim(A \to B) \leftrightarrow (A \to \sim B), \quad \sim(B \prec A) \leftrightarrow (\sim A \to \sim B)$$
$$\sim(A \to B) \leftrightarrow (\sim B \prec \sim A), \quad \sim(B \prec A) \leftrightarrow (\sim B \prec A).$$

Moreover, the following dual equivalences are *dually valid* in **2C**:

$$\sim\top \succ\!\!\prec \bot, \quad \sim\bot \succ\!\!\prec \top, \qquad \sim\sim A \succ\!\!\prec A$$
$$\sim(A \wedge B) \succ\!\!\prec (\sim A \vee \sim B), \quad \sim(A \vee B) \succ\!\!\prec (\sim A \wedge \sim B)$$
$$\sim(A \to B) \succ\!\!\prec (A \to \sim B), \quad \sim(B \prec A) \succ\!\!\prec (\sim A \to \sim B)$$
$$\sim(A \to B) \succ\!\!\prec (\sim B \prec \sim A), \quad \sim(B \prec A) \succ\!\!\prec (\sim B \prec A).$$

Hence, we may put $A' = \tau(A)$ or we may put $A' = \zeta(A)$ with the modification that for atomic formulas p, $\tau(\sim p) := \sim p$ and $\zeta(\sim p) := \sim p$.[5] $\qquad\square$

Lemma 1. *Let τ be the above mapping from $\mathcal{L}_{2C}$ into $\mathcal{L}_{\text{Int}}$, and let $\mathcal{M}' = \langle I, \leq, v^+, v^- \rangle$ be a model for **2C**. Consider the language $\mathcal{L}_{\text{Int}}$ based on the set of atoms $\Phi \cup \Phi'$, and let $\mathcal{M}$ be the structure $\langle I, \leq, v \rangle$, where the function v from $\Phi \cup \Phi'$ into subsets of I is defined by requiring for every $w \in I$ and $p \in \Phi$: $w \in v(p)$ iff $w \in v^+(p); w \in v(p')$ iff $w \in v^-(p)$. Clearly, $\mathcal{M}$ is a model for **Int**. For every $\mathcal{L}_{2C}$-formula A and every $w \in I$,*

1. $\mathcal{M}', w \models^+ A$ *iff* $\mathcal{M}, w \models^+ \tau(A)$,

---

[5]This modification has to be added in [25, proof of Corollary 4.4].

2. $\mathcal{M}', w \models^- A$ iff $\mathcal{M}, w \models^+ \tau(\sim A)$.

Proof. By simultaneous induction on the construction of A. We consider, by way of example, the case $A \equiv (B \prec C)$. Claim 1: $\mathcal{M}', x \models^+ (B \prec C)$ iff for every $x' \geq x$: $\mathcal{M}', x' \not\models^- C$ or $\mathcal{M}', x' \models^+ B$. The latter is the case iff for every $x' \geq x$: $\mathcal{M}, x' \not\models^+ \tau(\sim C)$ (by the induction hypothesis for 2.) or $\mathcal{M}, x' \models^+ \tau(B)$ (by the induction hypothesis for 1.), which holds iff $\mathcal{M}, x \models^+ \tau(\sim C) \rightarrow \tau(B)$ iff $\mathcal{M}, x \models^+ \tau(B \prec C)$. Claim 2: $\mathcal{M}', x \models^- (B \prec C)$ iff for every $x' \geq x$: $\mathcal{M}', x' \not\models^- C$ or $\mathcal{M}', x' \models^- B$. The latter is the case iff for every $x' \geq x$: $\mathcal{M}, x' \not\models^+ \tau(\sim C)$ (by the induction hypothesis for 2.) or $\mathcal{M}, x' \models^+ \tau(\sim B)$ (by the induction hypothesis for 2.), which holds iff $\mathcal{M}, x \models^+ \tau(\sim C) \rightarrow \tau(\sim B)$ iff $\mathcal{M}, x \models^+ \tau(\sim(B \prec C))$. $\qquad \square$

Lemma 2. Let again τ be the above mapping from $\mathcal{L}_{2C}$ into $\mathcal{L}_{Int}$, and let $\mathcal{M} = \langle I, \leq, v \rangle$ be a model for **Int**. Consider the language $\mathcal{L}_{Int}$ based on the set of atomic formulas $\Phi \cup \Phi'$, and let $\mathcal{M}'$ be the structure $\langle I, \leq, v^+, v^- \rangle$, where the mappings v^+, v^- from Φ into subsets of I are defined by requiring for every $w \in I$, every $p \in \Phi$: $w \in v(p)$ iff $w \in v^+(p)$; $w \in v(p')$ iff $w \in v^-(p)$. Clearly, $\mathcal{M}'$ is a model for **2C**. For every $\mathcal{L}_{2C}$-formula A and every $w \in I$,

1. $\mathcal{M}', w \models^+ A$ iff $\mathcal{M}, w \models^+ \tau(A)$,

2. $\mathcal{M}', w \models^- A$ iff $\mathcal{M}, w \models^+ \tau(\sim A)$.

Proof. Analogous to the proof of the previous lemma. $\qquad \square$

Theorem 1. For every $\mathcal{L}_{2C}$-formula A, $\models_{2C} A$ iff $\models_{Int} \tau(A)$.

Proof. Right to left: If $\not\models_{2C} A$, then there is a model $\mathcal{M}' = \langle I, \leq, v^+, v^- \rangle$ for **2C** and a state $w \in I$ with $\mathcal{M}', w \not\models^+ A$. By Lemma 1, in the model $\mathcal{M}$ for **Int** obtained from $\mathcal{M}'$ it holds that $\mathcal{M}, w \not\models^+ \tau(A)$. Left to right: If $\not\models_{Int} \tau(A)$, then there is a model $\mathcal{M} = \langle I, \leq, v \rangle$ for **Int** and a state $w \in I$ with $\mathcal{M}, w \not\models^+ \tau(A)$. By Lemma 2, in the model $\mathcal{M}'$ for **2C** obtained from $\mathcal{M}$ it holds that $\mathcal{M}', w \not\models^+ A$. $\qquad \square$

Lemma 3. Let ζ be the above mapping from $\mathcal{L}_{2C}$ into $\mathcal{L}_{DualInt}$ and, let $\mathcal{M}' = \langle I, \leq, v^+, v^- \rangle$ be a model for **2C**. Consider the language $\mathcal{L}_{DualInt}$ based on the set of atoms $\Phi \cup \Phi'$, and let $\mathcal{M}$ be the structure $\langle I, \leq, v \rangle$, where the function v from $\Phi \cup \Phi'$ into subsets of I is defined by requiring for every $w \in I$, every $p \in \Phi$: $w \in v(p)$ iff $w \in v^-(p)$; $w \in v(p')$ iff $w \in v^+(p)$. Clearly, $\mathcal{M}$ is a model for **DualInt**. For every $\mathcal{L}_{2C}$-formula A and every $w \in I$,

1. $\mathcal{M}', w \models^- A$ iff $\mathcal{M}, w \models^- \zeta(A)$,

2. $\mathcal{M}', w \models^+ A$ iff $\mathcal{M}, w \models^- \zeta(\sim A)$.

Moreover, López-Escobar assumed that a proof of $\sim A$ is a refutation of A (and not a proof of $(A \rightarrow \perp)$). Let $pr(\Pi, A)$ stand for "Π is a proof of A". Another fundamental assumption made in [10] is that $\{A \mid \exists \Pi, pr(\Pi, A) \text{ and } pr(\Pi, \sim A)\} = \varnothing$.

A formula A is said to be valid iff there exists a construction that is a proof of A, and it would make sense to consider co-implication as well and the dual validity of a formula A, understood as the existence of a construction that is a disproof of A. If we the stronger assumption that $\{A \mid \exists \Pi_1 \exists \Pi_2, pr(\Pi_1, A) \text{ and } pr(\Pi_2, \sim A)\} = \varnothing$ is made, then the rule *ex contradictione quodlibet* becomes validity preserving.

This proof-disproof interpretation is a semantics with respect to which David Nelson's constructive propositional logic with strong negation **N4** (or **N3**, if the above stronger assumption is made) is sound. Every theorem of **N4** is valid. On the proof-disproof interpretation *every* construction that proves or disproves a formula is both a proof and a disproof (refutation):

A construction c is a disproof of A iff c is a proof of $\sim A$.

A construction c is a proof of A iff c is a disproof of $\sim A$.

A so-called formulas-as-types notion of construction, i.e., an encoding of formal derivations by typed λ-terms, facilitates a denotational semantics of derivations through the denotational semantics of the typed λ-calculus, see, for example, [5]. The introduction and elimination rules for strongly negated formulas in natural deduction proof systems for **N4** and **N3** (cf. [7, 8]) suggest a typed λ-calculus in which *every* term that encodes a proof or a disproof occurs in more than one type. On the basis of López-Escobar's proposal, in [26, 28] a formulas-as-types notion of construction for substructural subsystems of Nelson's **N4** is developed that has the feature of non-unique typedness of terms. Instead of using terms of different sorts, it is postulated that the set of typed terms is the smallest set Γ that contains all typed variables and satisfies in addition to the familiar term-formation conditions the following equivalences:

- $M^{\sim \sim A} \in \Gamma$ iff $M^A \in \Gamma$;
- $M^{\sim (A \wedge B)} \in \Gamma$ iff $M^{(\sim A \vee \sim B)} \in \Gamma$;
- $M^{\sim (A \vee B)} \in \Gamma$ iff $M^{(\sim A \wedge \sim B)} \in \Gamma$;
- $M^{\sim (A \rightarrow B)} \in \Gamma$ iff $M^{(A \wedge \sim B)} \in \Gamma$.

As a result of requiring the satisfaction of theses equivalences, every term occurs in many types; the term $\langle x_1^A, x_2^{\sim B} \rangle$, for example, belongs to the types $(A \wedge \sim B)$, $\sim(A \rightarrow B)$, $\sim\sim(A \wedge \sim B)$, $\sim\sim\sim(A \rightarrow B)$, etc. In order to encode proofs and dual proofs in **2C**, we proceed differently.

Definition 8. The set of type symbols (or just types) is the set of all formulas of $\mathcal{L}_{2C}$. The set *Var* of term variables is defined as the union of two disjoint sets:

$$Var_p := \{x_i^{+A} \mid i \in \omega, A \text{ is a type}\} \text{ and } Var_{dp} := \{x_i^{-A} \mid i \in \omega, A \text{ is a type}\}.$$

We are thus considering *two sorts of typed variables*, those from Var_p, which we call proof variables (*p*-variables), and those from Var_{dp}, which we call dual proof variables (*dp*-variables). The variables are used to build up more complex typed terms, proof terms and dual proof terms. If M^{+A} $[M^{-A}]$ is a proof term [dual proof term] of type A, we also just write M^+ $[M^-]$ when the type is clear and no confusion is likely to arise, and sometimes we even just write M when it is clear whether M is a proof term or a dual proof term, and therefore no confusion is likely to arise.

Definition 9. We define the set of typed proofs terms (*p*-terms) $Term_p$ and the set of typed dual proof terms (*dp*-terms) $Term_{dp}$ by a simultaneous induction as the smallest sets Γ_p and Γ_{dp}, respectively, such that:

Pure clauses for *p*-terms

- $Var_p \subseteq \Gamma_p$;

- $\mathsf{Top}^{+\top} \in \Gamma_p$;

- if M^{+A}, $N^{+B} \in \Gamma_p$, then $\langle M^+, N^+ \rangle^{+(A \wedge B)} \in \Gamma_p$;

- if $M^{+A} \in \Gamma_p$, then $(K^0_{A,B} M^+)^{+(A \vee B)} \in \Gamma_p$;

- if $M^{+B} \in \Gamma_p$, then $(K^1_{A,B} M^+)^{+(A \vee B)} \in \Gamma_p$;

- if $M^{+B} \in \Gamma_p$, $x^{+A} \in Var_p$, then $(\lambda x.M^+)^{+(A \to B)} \in \Gamma_p$;

- if $M^{+\bot} \in \Gamma_p$ and A is a type, then $(ex_A M^+)^{+A} \in \Gamma_p$;

- if $M^{+(A \wedge B)} \in \Gamma_p$, then $(M^+)^{+A}_0$, $(M^+)^{+B}_1 \in \Gamma_p$;

- if x^{+A}, $y^{+B} \in Var_p$ and M^{+C}, N^{+C}, $G^{+(A \vee B)} \in \Gamma_p$, then $(K(x^+, M^+, y^+, N^+, G^+))^{+C} \in \Gamma_p$;

- if $M^{+(A \to B)}$, $N^{+A} \in \Gamma_p$, then $(M^+, N^+)^{+B} \in \Gamma_p$;

Pure clauses for *dp*-terms

- $Var_{dp} \subseteq \Gamma_{dp}$;

- $\mathsf{Bot}^{-\bot} \in \Gamma_{dp}$;

- if M^{-A}, $N^{-B} \in \Gamma_{dp}$, then $\langle M^-, N^- \rangle^{-(A \vee B)} \in \Gamma_{dp}$;

- if $M^{-A} \in \Gamma_{dp}$, then $(K^0_{A,B} M^-)^{(A \wedge B)} \in \Gamma_{dp}$;

- if $M^{-B} \in \Gamma_{dp}$, then $(K^1_{A,B} M^-)^{(A \wedge B)} \in \Gamma_{dp}$;

- if $M^{-B} \in \Gamma_p$, $x^{-A} \in Var_{dp}$, then $(\lambda x.M)^{-(B \prec A)} \in \Gamma_{dp}$;

- if $M^{-\top} \in \Gamma_{dp}$ and A is a type, then $(ex_A M^-)^{-A} \in \Gamma_{dp}$;

- if $M^{-(A \vee B)} \in \Gamma_{dp}$, then $(M^-)^{-A}_0$, $(M^-)^{-B}_1 \in \Gamma_{dp}$;

- if x^{-A}, $y^{-B} \in Var_{dp}$ and M^{-C}, N^{-C}, $G^{-(A \wedge B)} \in \Gamma_{dp}$, then $(K(x^-, M^-, y^-, N^-, G^-))^{-C} \in \Gamma_{dp}$;

- if $M^{-(B \prec A)}$, $N^{-A} \in \Gamma_{dp}$, then $(M^-, N^-)^{-B} \in \Gamma_{dp}$;

Mixed clauses

- if $M^{-B} \in \Gamma_{dp}$, $x^{+A} \in Var_p$, then $(\lambda x.M^-)^{-(A \to B)} \in \Gamma_{dp}$;

- if $M^{-(A \to B)} \in \Gamma_{dp}$, $N^{+A} \in \Gamma_p$, then $(M^-, N^+)^{-B} \in \Gamma_{dp}$;

- if $M^{+B} \in \Gamma_p$, $x^{-A} \in Var_{dp}$, then $(\lambda x.M^+)^{+(B \prec A)} \in \Gamma_p$;

- if $M^{+(B \prec A)} \in \Gamma_p$, $N^{-A} \in \Gamma_{dp}$, then $(M^+, N^-)^{+B} \in \Gamma_p$;

- if $M^{-A} \in \Gamma_{dp}$, then $(dp(M))^{+\sim A} \in \Gamma_p$;

- if $M^{+A} \in \Gamma_p$, then $(pd(M))^{-\sim A} \in \Gamma_{dp}$;

- if $M^{-\sim A} \in \Gamma_{dp}$, then $(dp(M))^{+A} \in \Gamma_p$;

- if $M^{+\sim A} \in \Gamma_p$, then $(pd(M))^{-A} \in \Gamma_{dp}$.

We define the set *Term* of typed terms as $Term_p \cup Term_{dp}$.[7]

Note that every $M \in Term$ is uniquely typed.

Definition 10. The set $fv(M)$ of free variables of $M \in Term$, is inductively defined as follows:

---

[7]Functional application, λ-abstraction, pairing, and left and right projection are well-known operations. The term constructors for disjunction, the left and right injections, are here denoted by '$K^0_{A,B}$' and '$K^1_{A,B}$' (for types A and B), and the term destructor for disjunction, the case analysis function, is here denoted by 'K'.

$fv(x) = \{x\}$, if $x \in \textit{Var}$;

$fv(\langle M, N \rangle) = fv(M) \cup fv(N)$;

$fv((M)_i) = fv(M)$, $i = 0, 1$;

$fv((K^i_{A,B} M)) = fv(M)$, $i = 0, 1$, for all types A, B;

$fv((\lambda x.M)) = fv(M) \setminus \{x\}$;

$fv((K(x, M, y, N, G))) = (fv(M) \cup fv(N) \cup fv(G)) \setminus \{x, y\}$;

$fv((M, N)) = fv(M) \cup fv(N)$;

$fv(\mathsf{Bot}^{-\perp}) = fv(\mathsf{Top}^{+\top}) = \varnothing$;

$fv(ex_A(M)) = fv(M)$, for every type A;

$fv(pd(M)) = fv(dp(M)) = fv(M)$.

A term variable is bound in term M (is an element of $bv(M)$) iff it does not belong to $fv(M)$.

We write $M \equiv N$ to express that M and N are the same symbols or are obtainable from each other by renaming bound variables. If $x \in \textit{Var}_p$ and $N \in \textit{Term}_p$, respectively $x \in \textit{Var}_{dp}$ and $N \in \textit{Term}_{pd}$ have the same type, then $M[x := N]$ is the result of substituting N for the occurrences of $x \in fv(M)$ in M.

Definition 11. The logical axiom-schema and rules of the two-sorted typed λ-calculus 2λ are:

$M = M$

If $M = N$, then $N = M$.

If $M = N$, $N = G$, then $M = G$.

If $M^{+(A \to B)} = N^{+(A \to B)}$, then $(M, G^{+A}) = (N, G^{+A})$.

If $M^{-(A \to B)} = N^{-(A \to B)}$, then $(M, G^{+A}) = (N, G^{+A})$.

If $M^{-(B \prec A)} = N^{-(B \prec A)}$, then $(M, G^{-A}) = (N, G^{-A})$.

If $M^{+(B \prec A)} = N^{+(B \prec A)}$, then $(M, G^{-A}) = (N, G^{-A})$.

If $M^{+A} = N^{+A}$, then $(G^{+(A \to B)}, M) = (G^{+(A \to B)}, N)$.

If $M^{+A} = N^{+A}$, then $(G^{-(A \to B)}, M) = (G^{-(A \to B)}, N)$.

If $M^{-A} = N^{-A}$, then $(G^{-(B \prec A)}, M) = (G^{-(B \prec A)}, N)$.

If $M^{-A} = N^{-A}$, then $(G^{+(B \prec A)}, M) = (G^{+(B \prec A)}, N)$.

If $M = N$, then $(\lambda x^{+A}.M) = (\lambda x^{+A}.N)$.

If $M = N$, then $(\lambda x^{-A}.M) = (\lambda x^{-A}.N)$.

The axiom-schemata of 2λ's theory of typed pd/dp-β-equality are:

$$((\lambda x^{+A}.M)N^{+A}) = M\,[x := N],\ ((\lambda x^{-A}.M)N^{-A}) = M\,[x := N],$$

$$dp(pd(M^{+A})) = M,\ pd(dp(M^{-A})) = M.$$

Definition 12. The binary relations $\longrightarrow_\beta$ (one-step pd/dp-β-reduction), $\longrightarrow\!\!\!\rightarrow_\beta$ (pd/dp-β-reduction), and $=_\beta$ (pd/dp-β-convertability) are defined as follows:

(1) $((\lambda x^{+A}.M)N^{+A}) \longrightarrow_\beta M\,[x := N]$; $((\lambda x^{-A}.M)N^{-A}) \longrightarrow_\beta M\,[x := N]$;

if $M^{+A} \longrightarrow_\beta N^{+A}$, then $(G^{+(A \to B)}, M) \longrightarrow_\beta (G^{+(A \to B)}, N)$;

if $M^{+A} \longrightarrow_\beta N^{+A}$, then $(G^{-(A \to B)}, M) \longrightarrow_\beta (G^{-(A \to B)}, N)$;

if $M^{-A} \longrightarrow_\beta N^{-A}$, then $(G^{-(B \prec A)}, M) \longrightarrow_\beta (G^{-(B \prec A)}, N)$;

if $M^{-A} \longrightarrow_\beta N^{-A}$, then $(G^{+(B \prec A)}, M) \longrightarrow_\beta (G^{+(B \prec A)}, N)$;

if $M^{+(A \to B)} \longrightarrow_\beta N^{+(A \to B)}$, then $(M, G^{+A}) \longrightarrow_\beta (N, G^{+A})$;

if $M^{-(A \to B)} \longrightarrow_\beta N^{-(A \to B)}$, then $(M, G^{+A}) \longrightarrow_\beta (N, G^{+A})$;

if $M^{-(B \prec A)} \longrightarrow_\beta N^{-(B \prec A)}$, then $(M, G^{-A}) \longrightarrow_\beta (N, G^{-A})$;

if $M^{+(B \prec A)} \longrightarrow_\beta N^{+(B \prec A)}$, then $(M, G^{-A}) \longrightarrow_\beta (N, G^{-A})$;

if $M \longrightarrow_\beta N$, then $(\lambda x^{+A}.M) \longrightarrow_\beta (\lambda x^{+A}.N)$;

if $M \longrightarrow_\beta N$, then $(\lambda x^{-A}.M) \longrightarrow_\beta (\lambda x^{-A}.N)$.

(2) $\longrightarrow\!\!\!\rightarrow_\beta$ is the reflexive and transitive closure of $\longrightarrow_\beta$;

(3) $=_\beta$ is the symmetric closure of $\longrightarrow\!\!\!\rightarrow_\beta$.

We show that every $M \in Term$ encodes a derivation in N2C, and that vice versa every derivation in N2C is encoded by some $M \in Term$.

Definition 13. If $P(\Pi, A, (\{A_1, \ldots, A_n\}; \{B_1, \ldots, B_m\}))$, then $M^{+A} \in Term_p$ is a construction of Π iff there are at most the variables $x_1^{+A_1}$, ..., $x_n^{+A_n}$, $x_1^{-B_1}$, ..., $x_m^{-B_m} \in fv(M)$. If $DP(\Pi, A, (\{A_1, \ldots, A_n\}; \{B_1, \ldots, B_m\}))$, then $M^{-A} \in Term_{dp}$ is a construction of Π iff there are at most the variables $x_1^{+A_1}$, ..., $x_n^{+A_n}$, $x_1^{-B_1}$, ..., $x_m^{-B_m} \in fv(M)$.

Proposition 5.

1. If $P(\Pi, A, (\Delta; \Gamma))$, then one can find a construction M^{+A} of Π.

   If $DP(\Pi, A, (\Delta; \Gamma))$, then one can find a construction M^{-A} of Π.

2. For every $M^{+A} \in Term_p$, one can find a proof of which M^{+A} is a construction. For every $M^{-A} \in Term_{dp}$, one can find a dual proof of which M^{-A} is a construction.

Proof. 1.: By a straightforward simultaneous induction on proofs and dual proofs. We consider some exemplary cases.

- If $P(\Pi, A, (\{A\}; \varnothing))$, we choose x^{+A}, and if $DP(\Pi, A, (\varnothing; \{A\}))$, we choose x^{-A}.

- If $P(\Pi, \top, (\varnothing; \varnothing))$, we choose $\mathsf{Top}^{+\top}$, and if $DP(\Pi, \bot, (\varnothing; \varnothing))$, we choose $\mathsf{Bot}^{-\bot}$.

- If $DP(\Pi, (A \rightarrow B), (\Delta; \Gamma))$ and if, by the induction hypothesis, $DP(\Pi', B, (\Delta \cup \{A\}; \Gamma))$ with N^{-B} being a construction of Π', then $(\lambda x^{+A}.N)^{-(A \rightarrow B)}$ is a construction of Π.

- If $DP(\Pi, B, (\Delta \cup \Gamma; \Delta' \cup \Gamma'))$ and if, by the induction hypothesis, $P(\Pi', A, (\Delta; \Gamma))$ and $DP(\Pi'', (A \rightarrow B), (\Delta'; \Gamma'))$ with N^{+A}, $G^{-(A \rightarrow B)}$ being constructions of Π' and Π'', respectively, then $(G, N)^{-B}$ is a construction of Π.

2.: By a straightforward simultaneous induction on proof terms and dual proof terms. Again, we consider some exemplary cases.

- If $M^{+A} \in Var_p$ or $M^{-A} \in Var_{dp}$, we choose A.

- If $M \equiv \mathsf{Top}^{+\top}$, we choose $\overline{\overline{\top}}$; if $M \equiv \mathsf{Bot}^{-\bot}$, we choose $\overline{\overline{\bot}}$.

- If $M \equiv (\lambda x^{-A}.N^{+B})^{+(B \prec A)}$, and, by the induction hypothesis, we have $P(\Pi, B, (\Delta; \Gamma))$, then M is a construction of $\overline{(B \prec A)}^{\Pi}$, where all counter-assumptions of A in Γ have been cancelled.

- If $M \equiv (N^{+(B \prec A)}, G^{-A})^{+B}$, and, by the induction hypotheses for proof terms and dual proof terms, $P(\Pi, (B \prec A), (\Delta; \Gamma))$ and $DP(\Pi', A, (\Delta'; \Gamma'))$, then M is a construction of $\overline{B}^{\Pi \ \Pi'}$. $\qquad \square$

5 Some peculiarities of 2C

The early systems of connexive logic have been criticized for their unintuitive or overly complicated semantics or for having some problematic properties, cf. [34], so that one may wonder about peculiarities of the system **2C**. First, note that like **C**, the system **2C** is a non-trivial inconsistent logic: For any A, $(A \wedge \sim A) \to \sim(A \wedge \sim A)$ and $\sim((A \wedge \sim A) \to \sim(A \wedge \sim A))$, for example, are both provable and, moreover, the formulas $\sim(A \vee \sim A) \prec (A \vee \sim A)$ and $\sim(\sim(A \vee \sim A) \prec (A \vee \sim A))$ are both dually provable. Also, $(A \wedge \sim A) \to A$ and $\sim((A \wedge \sim A) \to A)$ are both provable; this example is taken from [19].

Moreover, an anonymous referee pointed out that in **2C** for every formula A, the co-negation $-A$ of A is valid, the intuitionistic negation $\neg A$ of A is dually valid, the strong negation of $\neg A$ is valid, and the strong negation of $-A$ is dually valid. In terms of provability and dual provability we have the following derivations and encoding typed terms:

$$\frac{\dfrac{\overline{\top}}{\top \prec A}}{\sim(\top \prec A)}$$
$$(pd(\lambda x^{-A}.\mathsf{Top}^{+\top}))^{-\sim(\top \prec A)}$$

$$\frac{\dfrac{\overline{\underline{\bot}}}{A \to \bot}}{\sim(A \to \bot)}$$
$$(dp(\lambda x^{+A}.\mathsf{Bot}^{-\bot}))^{+\sim(A \to \bot)}$$

Is this peculiarity surprising, and is it a "pathology"? The provability of the formula $\sim(A \to \bot)$ shows that for every formula A, it is provably false that A implies absurdity (understood as a proposition the falsity of which is supported by every state). Since $\bot \to \bot$ is provable, we have another instance of inconsistency: $\bot \to \bot$ and $\sim(\bot \to \bot)$ are both provable. If the classical negation of a formula A is true in a classical model, then Aristotle's theses is not true for A in that model, so maybe it is not exceedingly surprising that the *connexive* logic **2C** validates $\sim(A \to \bot)$. I leave this issue for future discussion.

6 Summary and brief outlook

In this paper the bi-connexive logic **2C** has been motivated and introduced. Its natural deduction proof system N2C has been shown to be sound and complete for both provability and dual provability with respect to a certain class of relational models. It turned out that with respect to validity, **2C** is a conservative extension of and is faithfully embeddable into **Int**, and that with respect to dual validity, **2C** is a conservative extension of and is faithfully embeddable into **DualInt**. Moreover, it has been shown that proofs and dual proofs can be encoded by typed λ-terms from a two-sorted typed λ-calculus.

Topics for further investigation abound and include strong normalizability for N2C[8] and 2λ, a sequent calculus for **2C** and strong cut-elimination, the relation between normalization and cut-elimination, an extension of **2C** to first-order, the definition of other kinds of proof systems (tableaux, Fitch-style, ...), etc. Also, it would be interesting to characterize the set of $\mathcal{L}_{2C}$ formulas A for which it holds that A and $\sim A$ are both provable, respectively dually provable.

References

[1] A. Almukdad and D. Nelson. Constructible falsity and inexact predicates. *Journal of Symbolic Logic*, 49(1):231–233, 1984.

[2] J. M. Dunn. Partiality and its dual. *Studia Logica*, 66(1):5–40, 2000.

[3] S. Frankowski. Formalization of a plausible inference. *Bulletin of the Section of Logic*, 33(1):41–52, 2004.

[4] S. Frankowski. p-consequence versus q-consequence operations. *Bulletin of the Section of Logic*, 33(4):197–207, 2004.

[5] J.-Y. Girard. *Proofs and Types*. Cambridge University Press, Cambridge, 1989.

[6] R. Goré. Dual intuitionistic logic revisited. In R. Dyckhoff, editor, *Automated Reasoning with Analytic Tableaux and Related Methods*, volume 1847 of *Lecture Notes in AI*, pages 252–267. Springer Verlag, Berlin, 2000.

[7] N. Kamide and H. Wansing. Proof theory of Nelson's paraconsistent logic: A uniform perspective. *Theoretical Computer Science*, 415:1–38, 2012.

[8] N. Kamide and H. Wansing. *Proof Theory of N4-Related Paraconsistent Logics*. London, College Publications, 2015.

[9] N. Kamide and H. Wansing. *Completeness of connexive Heyting-Brouwer logic*. The IFCoLog Journal of Logics and Their Applications, 2016. this issue.

[10] E.G.K. López-Escobar. Refutability and elementary number theory. *Indigationes Mathematicae*, 75(4):362–374, 1972.

[11] G. Malinowski. Q-consequence operation. *Reports on Mathematical Logic*, 24:49–59, 1990.

[12] G. Malinowski. *Many-valued Logics*. Oxford, Clarendon Press, 1993.

[13] G. Malinowski. Beyond three inferential values. *Studia Logica*, 92(2):203–221, 2009.

[14] D. Nelson. Constructible falsity. *Journal of Symbolic Logic*, 14(1):16–26, 1949.

---

[8]A normal form theorem for the system **2Int**, of which **2C** is connexive variant, can be found in [35].

[15] D. Nelson. Negation and separation of concepts in constructive systems. In A. Heyting, editor, *Constructivity in Mathematics*, pages 208–225. North-Holland, Amsterdam, 1959.

[16] S. Odintsov. The class of extensions of Nelson's paraconsistent logic. *Studia Logica*, 80(2–3):291–320, 2005.

[17] S. Odintsov. *Constructive Negations and Paraconsistency*. Springer, Dordrecht, 2008.

[18] S. Odintsov and H. Wansing. The logic of generalized truth values and the logic of bilattices. *Studia Logica*, 103(1):91–112, 2015.

[19] H. Omori. A simple connexive extension of the basic relevant logic BD. *The IFCoLog Journal of Logics and their Applications*, 2016. this issue.

[20] C. Rauszer. An algebraic and Kripke-style approach to a certain extension of intuitionistic logic. *Dissertationes Mathematicae*, 167:1–62, 1980.

[21] Y. Shramko. Dual intuitionistic logic and a variety of negations: The logic of scientific research. *Studia Logica*, 80(2):347–367, 2005.

[22] Y. Shramko and H. Wansing. Some useful 16-valued logics: How a computer network should think. *Journal of Philosophical Logic*, 34(2):121–153, 2005.

[23] Y. Shramko and H. Wansing. *Truth and Falsehood: An Inquiry into Generalized Logical Values*. Springer, Dordrecht, 2011.

[24] D. van Dalen. Intuitionistic logic. In D. Gabbay and F. Guenthner, editors, *Handbook of Philosophical Logic*, volume III, pages 225–339. Reidel, Dordrecht, 1986.

[25] H. Wansing. Falsification, natural deduction and bi-intuitionistic logic. *Journal of Logic and Computation*, 26(2016):425–450. First published online July 17, 2013.

[26] H. Wansing. Formulas-as-types for a hierarchy of sublogics of intuitionistic propositional logic. In D. Pearce and H. Wansing, editors, *Non-classical Logics and Information Processing*, volume 619 of *Lecture Notes in AI*, pages 125–145. Springer, Berlin, 1992.

[27] H. Wansing. Informational interpretation of substructural propositional logics. *Journal of Logic, Language and Information*, 2(4):285–308, 1993.

[28] H. Wansing. *The Logic of Information Structures*, volume 681 of *Lecture Notes in AI*. Springer, Berlin, 1993.

[29] H. Wansing. Negation. In L. Goble, editor, *The Blackwell Guide to Philosophical Logic*, pages 415–436. Blackwell, Oxford, 2001.

[30] H. Wansing. Connexive modal logic. In R. Schmidt et al., editors, *Advances in Modal Logic*, pages 367–383. College Publications, London, 2005.

[31] H. Wansing. Contradiction and contrariety: Priest on negation. In J. Malinowski and A. Pietruszczak, editors, *Essays in Logic and Ontology*, volume 13 of *Poznań Studies in the Philosophy of the Sciences and the Humanities*, pages 81–93. Rodopi, Amsterdam, 2006.

[32] H. Wansing. Constructive negation, implication, and co-implication. *Journal of Applied Non-Classical Logics*, 18(2–3):341–364, 2008.

[33] H. Wansing. Proofs, disproofs, and their duals. In V. Goranko, L. Beklemishev, and V. Shehtman, editors, *Advances in Modal Logic*, volume 8, pages 483–505. College

Publications, London, 2010.

[34] H. Wansing. Connexive logic. In E. N. Zalta, editor, *The Stanford Encyclopedia of Philosophy*. Fall 2014 edition, 2014. `http://plato.stanford.edu/archives/fall2010/entries/logic-connexive/`.

[35] H. Wansing. A more general general proof theory. 2016, submitted.

[36] H. Wansing. On split negation, strong negation, information, falsification, and verification. In K. Bimbó, editor, *J. Michael Dunn on Information Based Logics*, volume 8 of *Outstanding Contributions to Logic*, pages 161–189. Springer, Dordrecht, 2016.

Received September 2015

COMPLETENESS OF CONNEXIVE HEYTING-BROUWER LOGIC

NORIHIRO KAMIDE

Department of Information and Electronic Engineering,
Teikyo University, Japan
`drnkamide08@kpd.biglobe.ne.jp`

HEINRICH WANSING

Department of Philosophy II,
Ruhr-University Bochum, Germany
`Heinrich.Wansing@rub.de`

Abstract

In this paper, we investigate a logic called *connexive Heyting-Brouwer logic* or *bi-intuitionistic connexive logic*, BCL. The system BCL is introduced as a Gentzen-type sequent calculus, and we prove some theorems for embedding BCL into a Gentzen-type sequent calculus BL for bi-intuitionistic logic, BiInt. The completeness theorem with respect to a Kripke semantics for BCL is proved using these embedding theorems. The cut-elimination theorem and a certain duality principle are also shown for some subsystems of BCL. Moreover, we present a sound and complete triply-signed tableau calculus for BCL.

1 Introduction

In [40], sixteen variants of *Heyting-Brouwer logic*, HB, also known as *bi-intuitionistic logic*, BiInt, are presented semantically and as display sequent systems. These logics differ in their treatment of strongly negated implications and co-implications. For want of a better terminology and notation they were referred to as systems (I_i, C_j), where i and j range over four ways of interpreting negated implications and co-implications, respectively.

We would like to thank the two anonymous referees and the participants of the *Symposium: Logic, Mathematics, and Perception*, (Stockholm, April 29, 2016), for their valuable comments. The first-mentioned author acknowledges support by JSPS KAKENHI Grant (C) 26330263 and by Grant-in-Aid for Okawa Foundation for Information and Telecommunications.

In this paper, we focus on the system (I_2, C_2) with a connexive reading of negated implications and co-implications. The system (I_2, C_2) is thus a *bi-intuitionistic connexive logic* (or *connexive Heyting-Brouwer logic*), and we therefore refer to it here as BCL. The logic BCL may also be seen as an extension of the connexive logic C from [39] by the co-implication of BiInt, presuming a connexive understanding of negated co-implications. Another understanding of co-implication is developed in [37, 43], and a natural deduction proof system and formulas-as-types notion of construction for a bi-connexive logic 2C that assumes this understanding of co-implication is presented in [42].

Systems of connexive logic and the bi-intuitionistic logic BiInt have been carefully studied since the 1960s and 1970s with various philosophical and mathematical motivations, see [2, 12, 13, 41] and [24, 25, 26, 9, 5]. The characteristic principles of connexive logic are usually traced back to Aristotle and Boethius, and the co-implication of BiInt can be traced back to Skolem [31] and Moisil, [15], see also [34].

A distinctive feature of connexive logics is that they validate the so-called

Aristotle's theses: $\sim(\alpha \to \sim\alpha)$ and $\sim(\sim\alpha \to \alpha)$, and
Boethius' theses: $(\alpha \to \beta) \to \sim(\alpha \to \sim\beta)$ and $(\alpha \to \sim\beta) \to \sim(\alpha \to \beta)$.

An intuitionistic (or constructive) connexive modal logic, CK, which is a constructive connexive analogue of the smallest normal modal logic K, was introduced in [39] by extending a certain basic intuitionistic (or constructive) connexive logic, C, which is a connexive variant of *Nelson's paraconsistent logic* [1, 16, 17, 8].[1] A *classical connexive modal logic* called CS4, which is based on the positive normal modal logic S4, was introduced in [7] as a Gentzen-type sequent calculus. The Kripke-completeness and cut-elimination theorems for CS4 were shown, and CS4 was shown to be embeddable into positive S4 and to be decidable. Moreover, it was shown in [7] that the basic constructive connexive logic C can be faithfully embedded into CS4 and into a subsystem of CS4 lacking syntactic duality between necessity and possibility.

Heyting-Brouwer logic, which is an extension of both *dual-intuitionistic logic*, DualInt, and intuitionistic logic, Int, was introduced by Rauszer [24, 25, 26], who proved algebraic and Kripke completeness theorems for BiInt. As was shown by Uustalu in 2003, cf.[20], the original Gentzen-type sequent calculus by Rauszer [24] does not enjoy cut-elimination, and various kinds of sequent systems for BiInt have been presented in the literature, including cut-free display sequent calculi in [5, 40], see also [21] and [20] for a comparison between sequent calculi for BiInt. Moreover, BiInt is known to be a logic that has a faithful embedding into the future-past tense

---

[1]Information on connexive logics can also be found on the web site [19].

logic KtT4 [10], and a modal logic based on BiInt was studied by Łukowski in [11].

Dual-intuitionistic logics are logics which have a Gentzen-type sequent calculus in which sequents have the restriction that the antecedent contains at most one formula [3, 4, 35]. This restriction of being singular in the antecedent is syntactically dual to that in Gentzen's sequent calculus LJ for intuitionistic logic, which is singular in the consequent. Historically speaking, the logics in the set of logics containing Czermak's *dual-intuitionistic calculus* [3], Goodman's *logic of contradiction* or *anti-intuitionistic logic* [4], and Urbas's extensions of Czermak's and Goodman's logics [35] were collectively referred to by Urbas as dual-intuitionistic logics. The dual-intuitionistic logic referred to as DualInt in [5, 37] is the implication-free fragment of BiInt (in a language with constants ⊥ and ⊤, but without intuitionistic negation as primitive). An interpretation of DualInt as the *logic of scientific research* was presented by Shramko in [29].

In this paper we combine the two approaches and introduce the *bi-intuitionistic connexive logic* (or *connexive Heyting-Brouwer logic*), BCL, as a Gentzen-type sequent calculus. The logic BCL may be seen as an extension of the connexive logic C from [39] by the co-implication of BiInt, using a connexive understanding of negated co-implications. Another understanding of co-implication is developed in [37, 43], and a natural deduction proof system and formulas-as-types notion of construction for a bi-connexive logic 2C that assumes this understanding of co-implication is presented in [42].

We will proceed as follows.

In Section 2, the logic BCL is introduced as a Gentzen-type sequent calculus, and a dual-valuation-style Kripke semantics for BCL is defined. BCL is constructed on the basis of Maehara's cut-free Gentzen-type sequent calculus LJ' for Int. We refer to a slightly modified version of LJ' here as IL. Gentzen-type sequent calculi ICL, DCL, BL, IL and DL for *intuitionistic connexive logic*, *dual-intuitionistic connexive logic*, *bi-intuitionistic logic*, *intuitionistic logic* and *dual-intuitionistic logic*, respectively, are defined as subsystems of BCL.

In Section 3, some theorems for syntactically and semantically embedding BCL into BL are proved, and using these theorems, the completeness theorem with respect to the Kripke semantics for BCL is shown as a central result of this paper. The cut-elimination theorems for ICL and DCL are shown using some restricted versions of the syntactical embedding theorem of BCL into BL. The cut-elimination theorem does *not* hold for BCL and BL.

In Section 4, some theorems for syntactically embedding ICL into DCL and vice versa are shown. These theorems reveal that ICL and DCL are syntactically dual to each other in a certain sense. Thus, it is shown in these theorems that BCL is constructed based on a duality principle of the characteristic subsystems.

Finally, in Section 5, we present a sound and complete tableau calculus for BCL and its subsystems ICL, DCL, BL, IL, and DL using triply-signed formulas.

2   Sequent calculus and Kripke semantics

Prior to a detailed discussion, we introduce the language of connexive Heyting-Brouwer logic, BCL. *Formulas* are constructed from countably many propositional variables $p, q, ...$, the binary connectives $\wedge$ (conjunction), $\vee$ (disjunction), $\rightarrow$ (implication), $\prec$ (co-implication), the constants $\bot$ and $\top$, and the unary $\sim$ (paraconsistent, strong negation). Greek small letters $\alpha, \beta, ...$ are used to denote formulas, and Greek capital letters $\Gamma, \Delta, ...$ are used to represent finite (possibly empty) sets of formulas. The symbol $\equiv$ is used to denote the equality of symbols. A *sequent* is an expression of the form $\Gamma \Rightarrow \Delta$. An expression $L \vdash \Gamma \Rightarrow \Delta$ means that $\Gamma \Rightarrow \Delta$ is provable in a sequent calculus L. A rule R of inference is said to be *admissible* in a sequent calculus L if the following condition is satisfied: For any instance

$$\frac{S_1 \cdots S_n}{S}$$

of R, if $L \vdash S_i$ for all i, then $L \vdash S$.

The *bi-intuitionistic connexive logic* BCL is introduced below as a Gentzen-type sequent calculus; BCL is based on Maehara's system LJ' for intuitionistic logic (see, e.g., p. 52 in [33]).

Definition 2.1 (BCL). The initial sequents of BCL are of the following form, for any propositional variable p:

$$p \Rightarrow p \qquad \sim p \Rightarrow \sim p$$

$$\Gamma \Rightarrow \Delta, \top \qquad \bot, \Gamma \Rightarrow \Delta \qquad \sim\top, \Gamma \Rightarrow \Delta \qquad \Gamma \Rightarrow \Delta, \sim\bot.$$

The structural inference rules of BCL are of the form:

$$\frac{\Gamma \Rightarrow \Delta, \alpha \quad \alpha, \Sigma \Rightarrow \Pi}{\Gamma, \Sigma \Rightarrow \Delta, \Pi} \text{ (cut)}$$

$$\frac{\Gamma \Rightarrow \Delta}{\alpha, \Gamma \Rightarrow \Delta} \text{ (we-left)} \qquad \frac{\Gamma \Rightarrow \Delta}{\Gamma \Rightarrow \Delta, \alpha} \text{ (we-right).}$$

The positive logical inference rules of BCL are of the form:

$$\frac{\Gamma \Rightarrow \Delta}{\top, \Gamma \Rightarrow \Delta} \text{ (}\top\text{-left)} \qquad \frac{\Gamma \Rightarrow \Delta}{\Gamma \Rightarrow \Delta, \bot} \text{ (}\bot\text{-right)}$$

$$\frac{\alpha, \beta, \Gamma \Rightarrow \Delta}{\alpha \wedge \beta, \Gamma \Rightarrow \Delta} \ (\wedge\text{left}) \qquad \frac{\Gamma \Rightarrow \Delta, \alpha \quad \Gamma \Rightarrow \Delta, \beta}{\Gamma \Rightarrow \Delta, \alpha \wedge \beta} \ (\wedge\text{right})$$

$$\frac{\alpha, \Gamma \Rightarrow \Delta \quad \beta, \Gamma \Rightarrow \Delta}{\alpha \vee \beta, \Gamma \Rightarrow \Delta} \ (\vee\text{left}) \qquad \frac{\Gamma \Rightarrow \Delta, \alpha, \beta}{\Gamma \Rightarrow \Delta, \alpha \vee \beta} \ (\vee\text{right})$$

$$\frac{\Gamma \Rightarrow \Delta, \alpha \quad \beta, \Sigma \Rightarrow \Pi}{\alpha{\rightarrow}\beta, \Gamma, \Sigma \Rightarrow \Delta, \Pi} \ (\rightarrow\text{left}) \qquad \frac{\alpha, \Gamma \Rightarrow \beta}{\Gamma \Rightarrow \alpha{\rightarrow}\beta} \ (\rightarrow\text{right})$$

$$\frac{\alpha \Rightarrow \Delta, \beta}{\alpha{\prec}\beta \Rightarrow \Delta} \ (\prec\text{left}) \qquad \frac{\Gamma \Rightarrow \Delta, \alpha \quad \beta, \Sigma \Rightarrow \Pi}{\Gamma, \Sigma \Rightarrow \Delta, \Pi, \alpha{\prec}\beta} \ (\prec\text{right}).$$

The negative logical inference rules of BCL are of the form:

$$\frac{\alpha, \Gamma \Rightarrow \Delta}{{\sim}{\sim}\alpha, \Gamma \Rightarrow \Delta} \ ({\sim}{\sim}\text{left}) \qquad \frac{\Gamma \Rightarrow \Delta, \alpha}{\Gamma \Rightarrow \Delta, {\sim}{\sim}\alpha} \ ({\sim}{\sim}\text{right})$$

$$\frac{\Gamma \Rightarrow \Delta}{{\sim}\bot, \Gamma \Rightarrow \Delta} \ ({\sim}\bot\text{-left}) \qquad \frac{\Gamma \Rightarrow \Delta}{\Gamma \Rightarrow \Delta, {\sim}\top} \ ({\sim}\top\text{-right})$$

$$\frac{{\sim}\alpha, \Gamma \Rightarrow \Delta \quad {\sim}\beta, \Gamma \Rightarrow \Delta}{{\sim}(\alpha \wedge \beta), \Gamma \Rightarrow \Delta} \ ({\sim}\wedge\text{ left}) \qquad \frac{\Gamma \Rightarrow \Delta, {\sim}\alpha, {\sim}\beta}{\Gamma \Rightarrow \Delta, {\sim}(\alpha \wedge \beta)} \ ({\sim}\wedge\text{ right})$$

$$\frac{{\sim}\alpha, {\sim}\beta, \Gamma \Rightarrow \Delta}{{\sim}(\alpha \vee \beta), \Gamma \Rightarrow \Delta} \ ({\sim}\vee\text{ left}) \qquad \frac{\Gamma \Rightarrow \Delta, {\sim}\alpha \quad \Gamma \Rightarrow \Delta, {\sim}\beta}{\Gamma \Rightarrow \Delta, {\sim}(\alpha \vee \beta)} \ ({\sim}\vee\text{ right})$$

$$\frac{\Gamma \Rightarrow \Delta, \alpha \quad {\sim}\beta, \Sigma \Rightarrow \Pi}{{\sim}(\alpha{\rightarrow}\beta), \Gamma, \Sigma \Rightarrow \Delta, \Pi} \ ({\sim}{\rightarrow}\text{left}) \qquad \frac{\alpha, \Gamma \Rightarrow {\sim}\beta}{\Gamma \Rightarrow {\sim}(\alpha{\rightarrow}\beta)} \ ({\sim}{\rightarrow}\text{right})$$

$$\frac{{\sim}\alpha \Rightarrow \Delta, \beta}{{\sim}(\alpha{\prec}\beta) \Rightarrow \Delta} \ ({\sim}{\prec}\text{ left}) \qquad \frac{\Gamma \Rightarrow \Delta, {\sim}\alpha \quad \beta, \Sigma \Rightarrow \Pi}{\Gamma, \Sigma \Rightarrow \Delta, \Pi, {\sim}(\alpha{\prec}\beta)} \ ({\sim}{\prec}\text{ right}).$$

Gentzen-type sequent calculi ICL, DCL, BL, IL and DL for *intuitionistic connexive logic*, *dual-intuitionistic connexive logic*, *bi-intuitionistic logic*, *intuitionistic logic* and *dual-intuitionistic logic*, respectively, are defined as subsystems of BCL.

Definition 2.2 (Subsystems of BCL).

1. ICL is the $\prec$-free part of BCL.

2. DCL is the $\rightarrow$-free part of BCL.

3. BL is the $\sim$-free part of BCL.

4. IL is the $\prec$-free part of BL.

5. DL is the $\rightarrow$-free part of BL.

We may note the following:

1. Let L be $L \in \{\text{BCL, ICL, DCL, BL, IL, DL}\}$. The sequents of the form $\alpha \Rightarrow \alpha$ for any formula α are provable in L. This fact can be shown by induction on α.

2. ($\rightarrow$right) and ($\prec$left) in BCL satisfy the single-succedent restriction and the single-antecedent restriction, respectively. These rules are the usual inference rules for the standard Gentzen-type sequent calculi LJ and DJ for intuitionistic logic and dual-intuitionistic logic, respectively. The same restrictions are also imposed to ($\sim\rightarrow$right) and ($\sim\prec$left) in BCL.

3. ($\sim\rightarrow$left), ($\sim\rightarrow$right), ($\sim\prec$left) and ($\sim\prec$right) correspond to the following characteristic axiom schemes for connexive logic:

   (a) $\sim(\alpha \rightarrow \beta) \leftrightarrow \alpha \rightarrow \sim\beta$,

   (b) $\sim(\alpha \prec \beta) \leftrightarrow \sim\alpha \prec \beta$.

4. A Gentzen-type sequent calculus LBiI for BiInt was presented in [20] based on Dragalin's sequent calculus for Int. LBiI has the logical inference rules of the form:

$$\frac{\alpha \rightarrow \beta, \Gamma \Rightarrow \Delta, \alpha \quad \beta, \Gamma \Rightarrow \Delta}{\alpha \rightarrow \beta, \Gamma \Rightarrow \Delta} \qquad \frac{\alpha, \Gamma \Rightarrow \beta}{\Gamma \Rightarrow \Delta, \alpha \rightarrow \beta}$$

$$\frac{\alpha \Rightarrow \Delta, \beta}{\alpha \prec \beta, \Gamma \Rightarrow \Delta} \qquad \frac{\Gamma \Rightarrow \Delta, \alpha \quad \beta, \Gamma \Rightarrow \Delta, \alpha \prec \beta}{\Gamma \Rightarrow \Delta, \alpha \prec \beta} \ .$$

It is known that the cut-elimination theorem does not hold for LBiI [20].

5. BL is theorem-equivalent to LBiI, and the cut-elimination theorem does not hold for BL and BCL. On the other hand, the cut-elimination theorem holds for ICL, DCL and DL.

6. A counterexample showing the failure of the cut-elimination in BL and BCL is presented as follows. This example is the same as that in [20]. The sequent $p \Rightarrow q, r \rightarrow ((p \prec q) \wedge r)$ where p, q and r are distinct propositional variables is

not provable in BL without (cut), but provable in BL with (cut) by:

$$
\begin{array}{c}
\dfrac{
\dfrac{p \Rightarrow p \quad q \Rightarrow q}{p \Rightarrow q, p \prec q}\ (\leftarrow\text{right})
\qquad
\dfrac{
\dfrac{
\begin{array}{cc}
\vdots & \vdots \\
p \prec q \Rightarrow p \prec q & r \Rightarrow r \\
\vdots & \vdots \\
\end{array}
}{
\dfrac{r, p \prec q, p \Rightarrow p \prec q \quad r, p \prec q, p \Rightarrow r}{r, p \prec q, p \Rightarrow (p \prec q) \wedge r}
}{p \prec q, p \Rightarrow r \rightarrow ((p \prec q) \wedge r)}\ (\rightarrow\text{right})
}{p, p \Rightarrow q, r \rightarrow ((p \prec q) \wedge r)}\ (\text{cut})
}{p \Rightarrow q, r \rightarrow ((p \prec q) \wedge r)}\ .
\end{array}
$$

7. It is known that the cut-elimination theorem holds for IL which is logically equivalent to a slightly modified version of Maehara's LJ′ for intuitionistic logic.

8. Intuitionistic negation is definable by $\neg_i \alpha := \alpha \rightarrow \bot$, and co-negation is definable by $\neg_d \alpha := \top \prec \alpha$. Moreover, in the presence of implication, $\top$ can be defined as $p \rightarrow p$, and in the presence of co-implication, $\bot$ can be defined as $p \prec p$, for some fixed propositional variable p.

Next, we introduce a Kripke semantics for BCL in which a distinction is drawn between positive valuations $\models^+$ and negative valuations $\models^-$.

Definition 2.3. A *Kripke frame* is a structure $\langle M, \leq \rangle$ satisfying the following conditions:

1. M is a nonempty set (of states),

2. $\leq$ is a preorder (i.e., a reflexive and transitive binary relation) on M.

Definition 2.4. *Connexive valuations* $\models^+$ and $\models^-$ on a Kripke frame $\langle M, \leq \rangle$ are mappings from the set Φ of propositional variables to the power set 2^M of M such that for any $\star \in \{+, -\}$, any $p \in \Phi$ and any $x, y \in M$, if $x \in \models^\star (p)$ and $x \leq y$, then $y \in \models^\star (p)$. We will write $x \models^\star p$ for $x \in \models^\star (p)$. These connexive valuations $\models^+$ and $\models^-$ are extended to mappings from the set of all formulas to 2^M by:

1. $x \models^+ \top$ always,

2. $x \models^+ \bot$ never,

3. $x \models^+ \alpha \wedge \beta$ iff $x \models^+ \alpha$ and $x \models^+ \beta$,

4. $x \models^+ \alpha \vee \beta$ iff $x \models^+ \alpha$ or $x \models^+ \beta$,

447

5. $x \models^+ \alpha \rightarrow \beta$ iff $\forall y \in M \ [x \leq y$ and $y \models^+ \alpha$ imply $y \models^+ \beta]$,

6. $x \models^+ \alpha \prec \beta$ iff $\exists y \in M \ [x \geq y$ and $y \models^+ \alpha$ and not-$(y \models^+ \beta)]$,

7. $x \models^+ \sim\alpha$ iff $x \models^- \alpha$,

8. $x \models^- \top$ never,

9. $x \models^- \bot$ always,

10. $x \models^- \alpha \wedge \beta$ iff $x \models^- \alpha$ or $x \models^- \beta$,

11. $x \models^- \alpha \vee \beta$ iff $x \models^- \alpha$ and $x \models^- \beta$,

12. $x \models^- \alpha \rightarrow \beta$ iff $\forall y \in M \ [x \leq y$ and $y \models^+ \alpha$ imply $y \models^- \beta]$,

13. $x \models^- \alpha \prec \beta$ iff $\exists y \in M \ [x \geq y$ and $y \models^- \alpha$ and not-$(y \models^+ \beta)]$,

14. $x \models^- \sim\alpha$ iff $x \models^+ \alpha$.

The following *hereditary condition* holds for $\models^+$ and $\models^-$: For any $\star \in \{+, -\}$, any formula α and any $x, y \in M$, if $x \models^\star \alpha$ and $x \leq y$, then $y \models^\star \alpha$.

Definition 2.5. A *connexive Kripke model* is a structure $\langle M, \leq, \models^+, \models^- \rangle$ such that

1. $\langle M, \leq \rangle$ is a Kripke frame,

2. $\models^+$ and $\models^-$ are connexive valuations on $\langle M, \leq \rangle$.

A formula α is *true* in a connexive Kripke model $\langle M, \leq, \models^+, \models^- \rangle$ if $x \models^+ \alpha$ for any $x \in M$. A formula α is *BCL-valid* in a Kripke frame $\langle M, \leq \rangle$ if it is true for all connexive valuations $\models^+$ and $\models^-$ on the Kripke frame. A formula α is *dually BCL-valid* in a Kripke frame $\langle M, \leq \rangle$ if for all connexive valuations $\models^+$ and $\models^-$ on the Kripke frame, $x \models^- \alpha$ for any $x \in M$. A set of formulas Γ *entails* a formula α in BCL ($\Gamma \models_{\text{BCL}} \alpha$) if whenever all formulas in Γ are BCL-valid in a frame, then so is α; Γ *dually entails* α in BCL ($\Gamma \models^d_{\text{BCL}} \alpha$) if whenever all formulas in Γ are dually BCL-valid in a frame, then so is α.

Obviously, in the presence of $\sim$, the notion of dual BCL-validity in a Kripke frame is definable in terms of BCL-validity. A formula α is dually BCL-valid in a Kripke frame iff $\sim\alpha$ is BCL-valid in that frame. In Section 5 we also consider a notion of dual provability, see also [37, 43].

Next, we present a Kripke semantics for BL. It has been emphasized in [40] that in the relational semantics of intuitionistic logic and Heyting-Brouwer logic, only verification conditions and no falsification conditions of formulas are specified. The reason why negative valuations $\models^-$ have not been considered in the literature on BiInt presumably is that its language lacks strong negation, $\sim$. Accordingly, the semantics of BL is presented in terms of ordinary Kripke models.

Definition 2.6. A *valuation* $\models$ on a Kripke frame $\langle M, \leq \rangle$ is a mapping from the set Φ of propositional variables to the power set 2^M of M such that for any $p \in \Phi$ and any $x, y \in M$, if $x \in \models (p)$ and $x \leq y$, then $y \in \models (p)$. We will write $x \models p$ for $x \in \models (p)$. This valuation $\models$ is extended to a mapping from the set of all formulas to 2^M by:

1. $x \models \top$ always,

2. $x \models \bot$ never,

3. $x \models \alpha \wedge \beta$ iff $x \models \alpha$ and $x \models \beta$,

4. $x \models \alpha \vee \beta$ iff $x \models \alpha$ or $x \models \beta$,

5. $x \models \alpha {\rightarrow} \beta$ iff $\forall y \in M \, [x \leq y$ and $y \models \alpha$ imply $y \models \beta]$,

6. $x \models \alpha {\prec} \beta$ iff $\exists y \in M \, [x \geq y$ and $y \models \alpha$ and not-$(y \models \beta)]$.[2]

The following hereditary condition holds for $\models$: For any formula α and any $x, y \in M$, if $x \models \alpha$ and $x \leq y$, then $y \models \alpha$.

Definition 2.7. A *Kripke model* is a structure $\langle M, \leq, \models \rangle$ such that

1. $\langle M, \leq \rangle$ is a Kripke frame,

2. $\models$ is a valuation on $\langle M, \leq \rangle$.

A formula α is *true* in a Kripke model $\langle M, \leq, \models \rangle$ if $x \models \alpha$ for any $x \in M$. A formula α is *BL-valid* in a Kripke frame $\langle M, \leq \rangle$ if it is true for every valuation $\models$ on the Kripke frame. A set of formulas Γ *entails* a formula α in BL ($\Gamma \models_{\text{BL}} \alpha$) if whenever all formulas in Γ are BL-valid in a frame, then so is α.

---

[2]One might, perhaps, expect from truth-conditions for implication and co-implication to be subject to "mutual duality" in the sense of obtainability from each other by interchanging between dual notions, such as $\forall$ and $\exists$, $\leq$ and $\geq$, "or" and "and" (with meta-negation being self dual). However, the conditions 3 and 4 in Definitions 2.4 and 2.6 seem not to be dual in this respect. Namely, the truth condition for co-implication should then be something like this:

$$x \models^+ \alpha {\prec} \beta \text{ iff } \exists y \in M \, [x \geq y \text{ and not-}(y \models^+ \alpha) \text{ and } y \models^+ \beta]$$

However, condition 4 in Definitions 2.4 and 2.6 is wide-spread in the literature; a discussion of this issue may be found, for example, in [28] and [30].

In addition one may define a formula α to be *dually BL-valid* in a Kripke frame $\langle M, \leq \rangle$ if for all connexive valuations $\models^+$ and $\models^-$ on the Kripke frame, $x \models^- \alpha$ for any $x \in M$ and say that Γ *dually entails* α in BL ($\Gamma \models^d_{\mathrm{BL}} \alpha$) if whenever all formulas in Γ are dually BL-valid in a Kripke frame, then so is α.

The following completeness theorem for bi-intuitionistic logic is well-known.

Proposition 2.8 (Completeness for BL). For any finite set of formulas $\Gamma \cup \{\alpha\}$, BL $\vdash \Gamma \Rightarrow \alpha$ iff $\Gamma \models_{\mathrm{BL}} \alpha$.

3   Embedding and completeness theorems

In the following, we introduce a translation of BCL into BL, and by using this translation, we show two theorems for syntactically and semantically embedding BCL into BL. A similar translation has been used by Gurevich [6], Rautenberg [27] and Vorob'ev [36] to embed Nelson's constructive logic N3 [16] into positive intuitionistic logic.

Definition 3.1. We fix a set Φ of propositional variables and define the set $\Phi' := \{p' \mid p \in \Phi\}$ of propositional variables. The language $\mathcal{L}_{\mathrm{BCL}}$ of BCL, introduced in the previous section, is based on Φ, $\top, \bot, \wedge, \vee, \rightarrow, \prec$ and $\sim$. The language $\mathcal{L}_{\mathrm{BL}}$ of BL is defined using Φ, Φ', $\top, \bot, \wedge, \vee, \rightarrow$ and $\prec$.

A mapping f from $\mathcal{L}_{\mathrm{BCL}}$ to $\mathcal{L}_{\mathrm{BL}}$ is defined inductively as follows.

1. for any $p \in \Phi$, $f(p) := p$ and $f(\sim p) := p' \in \Phi'$,

2. $f(\sharp) := \sharp$ with $\sharp \in \{\top, \bot\}$,

3. $f(\alpha \sharp \beta) := f(\alpha) \sharp f(\beta)$ with $\sharp \in \{\wedge, \vee, \rightarrow, \prec\}$,

4. $f(\sim\sim\alpha) := f(\alpha)$,

5. $f(\sim\top) := \bot$,

6. $f(\sim\bot) := \top$,

7. $f(\sim(\alpha \wedge \beta)) := f(\sim\alpha) \vee f(\sim\beta)$,

8. $f(\sim(\alpha \vee \beta)) := f(\sim\alpha) \wedge f(\sim\beta)$,

9. $f(\sim(\alpha \rightarrow \beta)) := f(\alpha) \rightarrow f(\sim\beta)$,

10. $f(\sim(\alpha \prec \beta)) := f(\sim\alpha) \prec f(\beta)$.

An expression $f(\Gamma)$ denotes the result of replacing every occurrence of a formula α in Γ by an occurrence of $f(\alpha)$. The same notation is used for the other mappings discussed later. Note that the function f is surjective but not injective.

Theorem 3.2 (Syntactical embedding from BCL into BL). Let Γ, Δ be finite sets of formulas in $\mathcal{L}_{\text{BCL}}$, and f be the mapping defined in Definition 3.1.

1. BCL $\vdash \Gamma \Rightarrow \Delta$ iff BL $\vdash f(\Gamma) \Rightarrow f(\Delta)$,

2. BCL $-$ (cut) $\vdash \Gamma \Rightarrow \Delta$ iff BL $-$ (cut) $\vdash f(\Gamma) \Rightarrow f(\Delta)$.

Proof. We show only (1) below.

- $(\Longrightarrow)$: By induction on the proofs P of $\Gamma \Rightarrow \Delta$ in BCL. We distinguish the cases according to the last inference of P, and show some cases.

1. Case $(\sim p \Rightarrow \sim p)$: The last inference of P is of the form: $\sim p \Rightarrow \sim p$ for any $p \in \Phi$. In this case, we obtain BL $\vdash f(\sim p) \Rightarrow f(\sim p)$, i.e., BL $\vdash p' \Rightarrow p'$ ($p' \in \Phi'$), by the definition of f.

2. Case $(\sim\!\to\text{right})$: The last inference of P is of the form:

$$\frac{\alpha, \Gamma \Rightarrow \sim\beta}{\Gamma \Rightarrow \sim(\alpha\!\to\!\beta)} \ (\sim\!\to\text{right}).$$

By induction hypothesis, we have BL $\vdash f(\alpha), f(\Gamma) \Rightarrow f(\sim\beta)$. Then, we obtain the required fact:

$$\frac{\vdots}{\frac{f(\alpha), f(\Gamma) \Rightarrow f(\sim\beta)}{f(\Gamma) \Rightarrow f(\alpha)\!\to\!f(\sim\beta)}} \ (\to\text{right})$$

where $f(\alpha)\!\to\!f(\sim\beta)$ coincides with $f(\sim(\alpha\!\to\!\beta))$ by the definition of f.

3. Case $(\sim\!\to\text{left})$: The last inference of P is of the form:

$$\frac{\Gamma \Rightarrow \Delta, \alpha \quad \sim\beta, \Sigma \Rightarrow \Pi}{\sim(\alpha\!\to\!\beta), \Gamma, \Sigma \Rightarrow \Delta, \Pi} \ (\sim\!\to\text{left}).$$

By induction hypothesis, we have BL $\vdash f(\Gamma) \Rightarrow f(\Delta), f(\alpha)$ and BL $\vdash f(\sim\beta), f(\Sigma) \Rightarrow f(\Pi)$. Then, we obtain the required fact:

$$\frac{\vdots \qquad\qquad \vdots}{\frac{f(\Gamma) \Rightarrow f(\Delta), f(\alpha) \quad f(\sim\beta), f(\Sigma) \Rightarrow f(\Pi)}{f(\alpha)\!\to\!f(\sim\beta), f(\Gamma), f(\Sigma) \Rightarrow f(\Delta), f(\Pi)}} \ (\to\text{left})$$

where $f(\alpha)\!\to\!f(\sim\beta)$ coincides with $f(\sim(\alpha\!\to\!\beta))$ by the definition of f.

4. Case ($\sim\sim$left): The last inference of P is of the form:

$$\frac{\alpha, \Gamma \Rightarrow \Delta}{\sim\sim\alpha, \Gamma \Rightarrow \Delta} \ (\sim\sim\text{left}).$$

By induction hypothesis, we have the required fact $\text{BL} \vdash f(\alpha), f(\Gamma) \Rightarrow f(\Delta)$ where $f(\alpha)$ coincides with $f(\sim\sim\alpha)$ by the definition of f.

5. Case (cut): The last inference of P is of the form:

$$\frac{\Gamma \Rightarrow \Delta, \alpha \quad \alpha, \Sigma \Rightarrow \Pi}{\Gamma, \Sigma \Rightarrow \Delta, \Pi} \ (\text{cut}).$$

By induction hypothesis, we have $\text{BL} \vdash f(\Gamma) \Rightarrow f(\Delta), f(\alpha)$ and $\text{BL} \vdash f(\alpha), f(\Sigma) \Rightarrow f(\Pi)$. Then, we obtain the required fact:

$$\frac{f(\Gamma) \Rightarrow \overset{\vdots}{f}(\Delta), f(\alpha) \quad f(\alpha), \overset{\vdots}{f}(\Sigma) \Rightarrow f(\Pi)}{f(\Gamma), f(\Sigma) \Rightarrow f(\Delta), f(\Pi)} \ (\text{cut}).$$

- ($\Longleftarrow$) : By induction on the proofs Q of $f(\Gamma) \Rightarrow f(\Delta)$ in BL. We distinguish the cases according to the last inference of Q, and show some cases.

1. Case ($\sim\sim$left): The last inference of Q is of the form:

$$\frac{f(\alpha), f(\Gamma) \Rightarrow f(\Delta)}{f(\sim\sim\alpha), f(\Gamma) \Rightarrow f(\Delta)} \ (\sim\sim\text{left})$$

where $f(\sim\sim\alpha)$ coincides with $f(\alpha)$ by the definition of f. By induction hypothesis, we have the required fact $\text{BCL} \vdash \alpha, \Gamma \Rightarrow \Delta$.

2. Case (cut): The last inference of Q is of the form:

$$\frac{f(\Gamma) \Rightarrow \overset{\vdots}{f}(\Delta), \beta \quad \beta, f(\Sigma) \overset{\vdots}{\Rightarrow} f(\Pi)}{f(\Gamma), f(\Sigma) \Rightarrow f(\Delta), f(\Pi)} \ (\text{cut}).$$

In this case, β is a formula of BL. We then have the fact $\gamma = f(\gamma)$ for any formula γ in BL. This can be shown by induction on γ. Thus, Q is of the form:

$$\frac{f(\Gamma) \Rightarrow \overset{\vdots}{f}(\Delta), f(\beta) \quad f(\beta), f(\Sigma) \overset{\vdots}{\Rightarrow} f(\Pi)}{f(\Gamma), f(\Sigma) \Rightarrow f(\Delta), f(\Pi)} \ (\text{cut}).$$

By induction hypothesis, we have BCL $\vdash \Gamma \Rightarrow \Delta, \beta$ and BCL $\vdash \beta, \Sigma \Rightarrow \Pi$. Then, we obtain the required fact:

$$
\frac{\Gamma \Rightarrow \overset{\vdots}{\Delta}, \beta \quad \beta, \overset{\vdots}{\Sigma} \Rightarrow \Pi}{\Gamma, \Sigma \Rightarrow \Delta, \Pi} \ (\text{cut}).
$$

3. Case (∧right): The last inference of Q is (∧right).

   (a) Subcase (1): The last inference of Q is of the form:

   $$
   \frac{f(\Gamma) \Rightarrow f(\Delta), f(\alpha) \quad f(\Gamma) \Rightarrow f(\Delta), f(\beta)}{f(\Gamma) \Rightarrow f(\Delta), f(\alpha \wedge \beta)} \ (\wedge \text{right})
   $$

   where $f(\alpha \wedge \beta)$ coincides with $f(\alpha) \wedge f(\beta)$ by the definition of f. By induction hypothesis, we have BCL $\vdash \Gamma \Rightarrow \Delta, \alpha$ and BCL $\vdash \Gamma \Rightarrow \Delta, \beta$. We thus obtain the required fact:

   $$
   \frac{\Gamma \Rightarrow \overset{\vdots}{\Delta}, \alpha \quad \Gamma \Rightarrow \overset{\vdots}{\Delta}, \beta}{\Gamma \Rightarrow \Delta, \alpha \wedge \beta} \ (\wedge \text{right}).
   $$

   (b) Subcase (2): The last inference of Q is of the form:

   $$
   \frac{f(\Gamma) \Rightarrow f(\Delta), f(\sim\alpha) \quad f(\Gamma) \Rightarrow f(\Delta), f(\sim\beta)}{f(\Gamma) \Rightarrow f(\Delta), f(\sim(\alpha \vee \beta))} \ (\wedge \text{right})
   $$

   where $f(\sim(\alpha \vee \beta))$ coincides with $f(\sim\alpha) \wedge f(\sim\beta)$ by the definition of f. By induction hypothesis, we have BCL $\vdash \Gamma \Rightarrow \Delta, \sim\alpha$ and BCL $\vdash \Gamma \Rightarrow \Delta, \sim\beta$. We thus obtain the required fact:

   $$
   \frac{\Gamma \Rightarrow \overset{\vdots}{\Delta}, \sim\alpha \quad \Gamma \Rightarrow \overset{\vdots}{\Delta}, \sim\beta}{\Gamma \Rightarrow \Delta, \sim(\alpha \vee \beta)} \ (\sim \vee \text{right}).
   $$

   $\square$

We can obtain a theorem for syntactically embedding ICL into IL.

Theorem 3.3 (Syntactical embedding from ICL into IL). Suppose that $\mathcal{L}_{\text{ICL}}$ and $\mathcal{L}_{\text{IL}}$ are obtained from $\mathcal{L}_{\text{BCL}}$ and $\mathcal{L}_{\text{BL}}$, respectively, by deleting $\prec$. Let Γ, Δ be finite sets of formulas in $\mathcal{L}_{\text{ICL}}$, and f be the mapping from $\mathcal{L}_{\text{ICL}}$ into $\mathcal{L}_{\text{IL}}$, which is obtained from Definition 3.1 by deleting the conditions of $\prec$.

1. $\text{ICL} \vdash \Gamma \Rightarrow \Delta$ iff $\text{IL} \vdash f(\Gamma) \Rightarrow f(\Delta)$,

2. $\text{ICL} - (\text{cut}) \vdash \Gamma \Rightarrow \Delta$ iff $\text{IL} - (\text{cut}) \vdash f(\Gamma) \Rightarrow f(\Delta)$.

We can also obtain a theorem for syntactically embedding DCL into DL.

Theorem 3.4 (Syntactical embedding from DCL into DL). Suppose that $\mathcal{L}_{\text{DCL}}$ and $\mathcal{L}_{\text{DL}}$ are obtained from $\mathcal{L}_{\text{BCL}}$ and $\mathcal{L}_{\text{BL}}$, respectively, by deleting $\to$. Let Γ, Δ be finite sets of formulas in $\mathcal{L}_{\text{DCL}}$, and f be the mapping from $\mathcal{L}_{\text{DCL}}$ into $\mathcal{L}_{\text{DL}}$, which is obtained from Definition 3.1 by deleting the conditions of $\to$.

1. $\text{DCL} \vdash \Gamma \Rightarrow \Delta$ iff $\text{DL} \vdash f(\Gamma) \Rightarrow f(\Delta)$,

2. $\text{DCL} - (\text{cut}) \vdash \Gamma \Rightarrow \Delta$ iff $\text{DL} - (\text{cut}) \vdash f(\Gamma) \Rightarrow f(\Delta)$.

Using Theorems 3.3 and 3.4, we can obtain the cut-elimination theorems for ICL and DCL.

Theorem 3.5 (Cut-elimination for ICL and DCL). Let L be ICL or DCL. The rule (cut) is admissible in cut-free L.

Proof. We show only the case for ICL. Suppose that $\text{ICL} \vdash \Gamma \Rightarrow \Delta$. Then, we have $\text{IL} \vdash f(\Gamma) \Rightarrow f(\Delta)$ by Theorem 3.3 (1), and hence $\text{IL} - (\text{cut}) \vdash f(\Gamma) \Rightarrow f(\Delta)$ by the cut-elimination theorem for IL (this is known [33]). By Theorem 3.3 (2), we obtain $\text{ICL} - (\text{cut}) \vdash \Gamma \Rightarrow \Delta$. $\qquad\qquad\square$

Next, we show a theorem for semantically embedding BCL into BL.

Lemma 3.6. Let f be the mapping defined in Definition 3.1. For any connexive Kripke model $\langle M, \leq, \models^+, \models^- \rangle$, we can construct a Kripke model $\langle M, \leq, \models \rangle$ such that for any formula α and any $x \in M$,

1. $x \models^+ \alpha$ iff $x \models f(\alpha)$,

2. $x \models^- \alpha$ iff $x \models f(\sim\alpha)$.

Proof. Let Φ be a set of propositional variables and Φ' be the set $\{p' \mid p \in \Phi\}$ of propositional variables. Suppose that $\langle M, \leq, \models^+, \models^- \rangle$ is a connexive Kripke model where $\models^+$ and $\models^-$ are mappings from Φ to the power set 2^M of M, and that the hereditary condition with respect to $p \in \Phi$ holds for $\models^+$ and $\models^-$. Suppose that $\langle M, \leq, \models \rangle$ is a Kripke model where $\models$ is a mapping from $\Phi \cup \Phi'$ to 2^M, and that the hereditary condition with respect to $p \in \Phi \cup \Phi'$ holds for $\models$. Suppose moreover that these models satisfy the following conditions: For any $x \in M$ and any $p \in \Phi$,

1. $x \models^+ p$ iff $x \models p$,

2. $x \models^{-} p$ iff $x \models p'$.

Then, the lemma is proved by (simultaneous) induction on the complexity of α.

- Base step:

  Case $\alpha \equiv p \in \Phi$: For (1), we obtain: $x \models^{+} p$ iff $x \models p$ iff $x \models f(p)$ (by the definition of f). For (2), we obtain: $x \models^{-} p$ iff $x \models p'$ iff $x \models f(\sim p)$ (by the definition of f).

- Induction step:

1. Case $\alpha \equiv \top$: For (1), we obtain: $x \models^{+} \top$ iff $x \models \top$ iff $x \models f(\top)$ (by the definition of f). For (2), we obtain: $x \models^{-} \top$ iff $x \models \bot$ iff $x \models f(\sim\top)$ (by the definition of f).

2. Case $\alpha \equiv \bot$: Similar to the above case.

3. Case $\alpha \equiv \beta \wedge \gamma$: For (1), we obtain: $x \models^{+} \beta \wedge \gamma$ iff $x \models^{+} \beta$ and $x \models^{+} \gamma$ iff $x \models f(\beta)$ and $x \models f(\gamma)$ (by induction hypothesis for 1) iff $x \models f(\beta) \wedge f(\gamma)$ iff $x \models f(\beta \wedge \gamma)$ (by the definition of f). For (2), we obtain: $x \models^{-} \beta \wedge \gamma$ iff $x \models^{-} \beta$ or $x \models^{-} \gamma$ iff $x \models f(\sim\beta)$ or $x \models f(\sim\gamma)$ (by induction hypothesis for 2) iff $x \models f(\sim\beta) \vee f(\sim\gamma)$ iff $x \models f(\sim(\beta \wedge \gamma))$ (by the definition of f).

4. Case $\alpha \equiv \beta \vee \gamma$: Similar to the above case.

5. Case $\alpha \equiv \beta \rightarrow \gamma$: For (1), we obtain: $x \models^{+} \beta \rightarrow \gamma$ iff $\forall y \in M \ [x \leq y$ and $y \models^{+} \beta$ imply $y \models^{+} \gamma]$ iff $\forall y \in M \ [x \leq y$ and $y \models f(\beta)$ imply $y \models f(\gamma)]$ (by induction hypothesis for 1) iff $x \models f(\beta) \rightarrow f(\gamma)$ iff $x \models f(\beta \rightarrow \gamma)$ (by the definition of f). For (2), we obtain: $x \models^{-} \beta \rightarrow \gamma$ iff $\forall y \in M \ [x \leq y$ and $y \models^{+} \beta$ imply $y \models^{-} \gamma]$ iff $\forall y \in M \ [x \leq y$ and $y \models f(\beta)$ imply $y \models f(\sim\gamma)]$ (by induction hypotheses for 1 and 2) iff $x \models f(\beta) \rightarrow f(\sim\gamma)$ iff $x \models f(\sim(\beta \rightarrow \gamma))$ (by the definition of f).

6. Case $\alpha \equiv \beta \prec \gamma$: For (1), we obtain: $x \models^{+} \beta \prec \gamma$ iff $\exists y \in M \ [y \leq x$ and $y \models^{+} \beta$ and not-$(y \models^{+} \gamma)]$ iff $\exists y \in M \ [y \leq x$ and $y \models f(\beta)$ and not-$(y \models f(\gamma))]$ (by induction hypothesis for 1) iff $x \models f(\beta) \prec f(\gamma)$ iff $x \models f(\beta \prec \gamma)$ (by the definition of f). For (2), we obtain: $x \models^{-} \beta \prec \gamma$ iff $\exists y \in M \ [y \leq x$ and $y \models^{-} \beta$ and not-$(y \models^{+} \gamma)]$ iff $\exists y \in M \ [y \leq x$ and $y \models f(\sim\beta)$ and not-$(y \models f(\gamma))]$ (by induction hypotheses for 1 and 2) iff $x \models f(\sim\beta) \prec f(\gamma)$ iff $x \models f(\sim(\beta \prec \gamma))$ (by the definition of f).

7. Case $\alpha \equiv \sim\beta$: For (1), we obtain: $x \models^{+} \sim\beta$ iff $x \models^{-} \beta$ iff $x \models f(\sim\beta)$ (by induction hypothesis for 2). For (2), we obtain: $x \models^{-} \sim\beta$ iff $x \models^{+} \beta$ iff $x \models f(\beta)$ (by induction hypothesis for 1) iff $x \models f(\sim\sim\beta)$ (by the definition of f). $\square$

Lemma 3.7. Let f be the mapping defined in Definition 3.1. For any Kripke model $\langle M, \leq, \models \rangle$, we can construct a connexive Kripke model $\langle M, \leq, \models^+, \models^- \rangle$ such that for any formula α and any $x \in M$,

1. $x \models f(\alpha)$ iff $x \models^+ \alpha$,

2. $x \models f(\sim\alpha)$ iff $x \models^- \alpha$.

Proof. Similar to the proof of Lemma 3.6. $\qquad\qquad\qquad\qquad\qquad\qquad\qquad$ $\square$

Theorem 3.8 (Semantical embedding from BCL into BL). Let f be the mapping defined in Definition 3.1. For any any finite set of formula $\Gamma \cup \{\alpha\}$, $f(\Gamma) \models_{\mathrm{BL}} f(\alpha)$ iff $\Gamma \models_{\mathrm{BCL}} \alpha$.

Proof. By Lemmas 3.6 and 3.7. $\qquad\qquad\qquad\qquad\qquad\qquad\qquad\qquad\qquad$ $\square$

Theorem 3.9 (Completeness for BCL). For any finite set of formula $\Gamma \cup \{\alpha\}$, BCL $\vdash \Gamma \Rightarrow \alpha$ iff $\Gamma \models_{\mathrm{BCL}} \alpha$ (and thus also BCL $\vdash \sim\Gamma \Rightarrow \sim\alpha$ iff $\Gamma \models^d_{\mathrm{BCL}} \alpha$).

Proof. BCL $\vdash \Gamma \Rightarrow \alpha$ iff BL $\vdash f(\Gamma) \Rightarrow f(\alpha)$ (by Theorem 3.2) iff $f(\Gamma) \models_{\mathrm{BL}} f(\alpha)$ (by Proposition 2.8) iff $\Gamma \models_{\mathrm{BCL}} \alpha$ (by Theorem 3.8). $\qquad\qquad\qquad$ $\square$

4 Duality between subsystems

We show that ICL and DCL can be syntactically embedded into each other. These results show that BCL is constructed based on a duality between ICL and DCL. Firstly, we introduce a translation from ICL into DCL. The idea of this translation comes from [3, 35].

Definition 4.1. We fix a common set Φ of propositional variables. The language $\mathcal{L}_{\mathrm{ICL}}$ of ICL is defined using Φ, $\bot, \wedge, \vee, \rightarrow$ and $\sim$. The language $\mathcal{L}_{\mathrm{DCL}}$ of DCL is defined using Φ, $\top, \wedge, \vee, \prec$ and $\sim$.

A mapping f from $\mathcal{L}_{\mathrm{ICL}}$ to $\mathcal{L}_{\mathrm{DCL}}$ is defined inductively as follows.

1. $f(p) := p$ for any $p \in \Phi$,

2. $f(\bot) := \top$,

3. $f(\alpha \wedge \beta) := f(\alpha) \vee f(\beta)$,

4. $f(\alpha \vee \beta) := f(\alpha) \wedge f(\beta)$,

5. $f(\alpha \rightarrow \beta) := f(\beta) \prec f(\alpha)$,

6. $f(\sim\alpha) := \sim f(\alpha)$.

We then obtain a theorem for syntactically embedding ICL into DCL.

Theorem 4.2 (Syntactical embedding from ICL into DCL). Let Γ and Δ be finite sets of formulas in $\mathcal{L}_{\mathrm{ICL}}$, and f be the mapping defined in Definition 4.1.

1. ICL $\vdash \Gamma \Rightarrow \Delta$ iff DCL $\vdash f(\Delta) \Rightarrow f(\Gamma)$,

2. ICL $-$ (cut) $\vdash \Gamma \Rightarrow \Delta$ iff DCL $-$ (cut) $\vdash f(\Delta) \Rightarrow f(\Gamma)$.

Proof. We show only the direction ($\Longrightarrow$) of (1) by induction on the proofs P of $\Gamma \Rightarrow \Delta$ in ICL. We distinguish the cases according to the last inference of P, and show some cases.

1. Case ($\sim\rightarrow$left): The last inference of P is of the form:

$$\frac{\Gamma \Rightarrow \Delta, \alpha \quad \sim\beta, \Sigma \Rightarrow \Pi}{\sim(\alpha\rightarrow\beta), \Gamma, \Sigma \Rightarrow \Delta, \Pi} \ (\sim\rightarrow\text{left}).$$

By induction hypothesis, we have DCL $\vdash f(\alpha), f(\Delta) \Rightarrow f(\Gamma)$ and DCL $\vdash f(\Pi) \Rightarrow f(\Sigma), f(\sim\beta)$ where $f(\sim\beta)$ coincides with $\sim f(\beta)$ by the definition of f. Then, we obtain the required fact:

$$\frac{\begin{array}{cc} \vdots & \vdots \\ f(\Pi) \Rightarrow f(\Sigma), \sim f(\beta) & f(\alpha), f(\Delta) \Rightarrow f(\Gamma) \end{array}}{f(\Pi), f(\Delta) \Rightarrow f(\Sigma), f(\Gamma), \sim(f(\beta)\prec f(\alpha))} \ (\sim\!\!\prec\text{right})$$

where $\sim(f(\beta)\prec f(\alpha))$ coincides with $f(\sim(\alpha\rightarrow\beta))$ by the definition of f.

2. Case ($\sim\rightarrow$right): The last inference of P is of the form:

$$\frac{\alpha, \Gamma \Rightarrow \sim\beta}{\Gamma \Rightarrow \sim(\alpha\rightarrow\beta)} \ (\sim\rightarrow\text{right}).$$

By induction hypothesis, we have DCL $\vdash f(\sim\beta) \Rightarrow f(\Gamma), f(\alpha)$ where $f(\sim\beta)$ coincides with $\sim f(\beta)$ by the definition of f. Then, we obtain the required fact:

$$\frac{\begin{array}{c} \vdots \\ \sim f(\beta) \Rightarrow f(\Gamma), f(\alpha) \end{array}}{\sim(f(\beta)\prec f(\alpha)) \Rightarrow f(\Gamma)} \ (\sim\!\!\prec\text{left})$$

where $\sim(f(\beta)\prec f(\alpha))$ coincides with $f(\sim(\alpha\rightarrow\beta))$ by the definition of f. $\quad\square$

Similarly, we can introduce a translation from DCL into ICL.

Definition 4.3. Φ, $\mathcal{L}_{\mathrm{DCL}}$ and $\mathcal{L}_{\mathrm{ICL}}$ are the same as in Definition 4.1. A mapping g from $\mathcal{L}_{\mathrm{DCL}}$ to $\mathcal{L}_{\mathrm{ICL}}$ is defined inductively as follows.

1. $g(p) := p$ for any $p \in \Phi$,

2. $g(\top) := \bot$,

3. $g(\alpha \wedge \beta) := g(\alpha) \vee g(\beta)$,

4. $g(\alpha \vee \beta) := g(\alpha) \wedge g(\beta)$,

5. $g(\alpha \prec \beta) := g(\beta) {\rightarrow} g(\alpha)$,

6. $g(\sim\alpha) := \sim g(\alpha)$.

We can obtain a theorem for syntactically embedding DCL into ICL.

Theorem 4.4 (Syntactical embedding from DCL into ICL). Let Γ and Δ be finite sets of formulas in $\mathcal{L}_{\mathrm{DCL}}$ and g be the mapping defined in Definition 4.3.

1. $\mathrm{DCL} \vdash \Gamma \Rightarrow \Delta$ iff $\mathrm{ICL} \vdash g(\Delta) \Rightarrow g(\Delta)$,

2. $\mathrm{DCL} - (\mathrm{cut}) \vdash \Gamma \Rightarrow \Delta$ iff $\mathrm{ICL} - (\mathrm{cut}) \vdash g(\Delta) \Rightarrow g(\Delta)$.

Proof. Similar to Theorem 4.2. □

Note that the following holds for ICL and DCL:

1. $\mathrm{ICL} \vdash gf(\Gamma) \Rightarrow gf(\Delta)$ iff $\mathrm{ICL} \vdash \Gamma \Rightarrow \Delta$,

2. $\mathrm{DCL} \vdash fg(\Gamma) \Rightarrow fg(\Delta)$ iff $\mathrm{DCL} \vdash \Gamma \Rightarrow \Delta$.

Similarly, we can introduce translations from IL into DL and vice versa, and can show the syntactical embedding theorems based on these translations.

Using Theorems 4.2, 4.4 and 3.5, we can obtain alternative proofs of the cut-elimination theorems for ICL, DCL, IL and DL.

5 Tableau calculus

A sound and complete tableau calculus for connexive Heyting-Brouwer logic can be obtained by modifying the tableau calculus for the modal logic BS4 from [18], which was obtained by modifying Priest's [22, 23] tableau calculus for the modal logic

S4$_{\text{FDE}}$ (or K$_{\text{FDE}\rho\tau}$ as Priest calls the latter system), i.e., S4 based on first-degree entailment logic.

We assume some familiarity with the tableau method as applied by Priest. We define tableau calculi for BCL and its subsystems ICL, DCL, BL, IL, and DL from Definition 2.2. Since the languages of BL, IL, and DL lack strong negation, support of falsity for a formula α cannot be captured as the support of truth of $\sim\alpha$, and the tableau nodes have to provide information concerning:

- support of truth, indicated by $+T$,

- failure to support truth, indicated by $+F$,

- support of falsity, indicated by $-T$,

- failure to support falsity, indicated by $-F$.[3]

Moreover, the nodes have to provide information about "accessibility" between states. Accordingly, in tableaux for BCL the tableau entries are of the form $\alpha, +Ti$, or $\alpha, +Fi$, or $\alpha, -Ti$, or $\alpha, -Fi$, or irj, where α is a formula from $\mathcal{L}_{\text{BCL}}$, i and j are natural numbers representing states, and irj is to be understood as $i \leq j$. We distinguish between single conclusion derivability statement $\Delta \vdash \beta$ and single conclusion dual derivability statement $\Delta \vdash^d \beta$. Tableaux for a single conclusion derivability statement $\Delta \vdash \beta$ start with nodes of the form $\alpha, +T0$ for every premise α from the finite premise set Δ and a node of the form $\beta, +F0$. Tableaux for a single conclusion dual derivability statement $\Delta \vdash^d \beta$ start with nodes of the form $\alpha, -T0$ for every premise α from the finite premise set Δ and a node of the form $\beta, -F0$.

Tableau rules are applied to tableau nodes, thereby leading to more complex, expanded tableaux. A branch of a tableau closes iff it contains a pair of nodes $\alpha, +Ti$ and $\alpha, +Fi$ or a pair of nodes $\alpha, -Ti$ and $\alpha, -Fi$. The tableau closes iff all of its branches close. If a tableau (tableau branch) is not closed, it is called open. A tableau branch is said to be completed iff no more rules can be applied to expand it. A tableau is said to be completed iff each of its branches is completed.

Definition 5.1. The triply-signed tableau calculus for BCL consists of the following rules:

---

[3]An anonymous referee proposed to let $+F$ stand for support of falsity and $-T$ for failure of a truth support. This is certainly suggestive, but on the other hand the signs '+' and '−' remind one of the support of truth and support of falsity relations, so that we prefer to use the above notation.

Decomposition rules

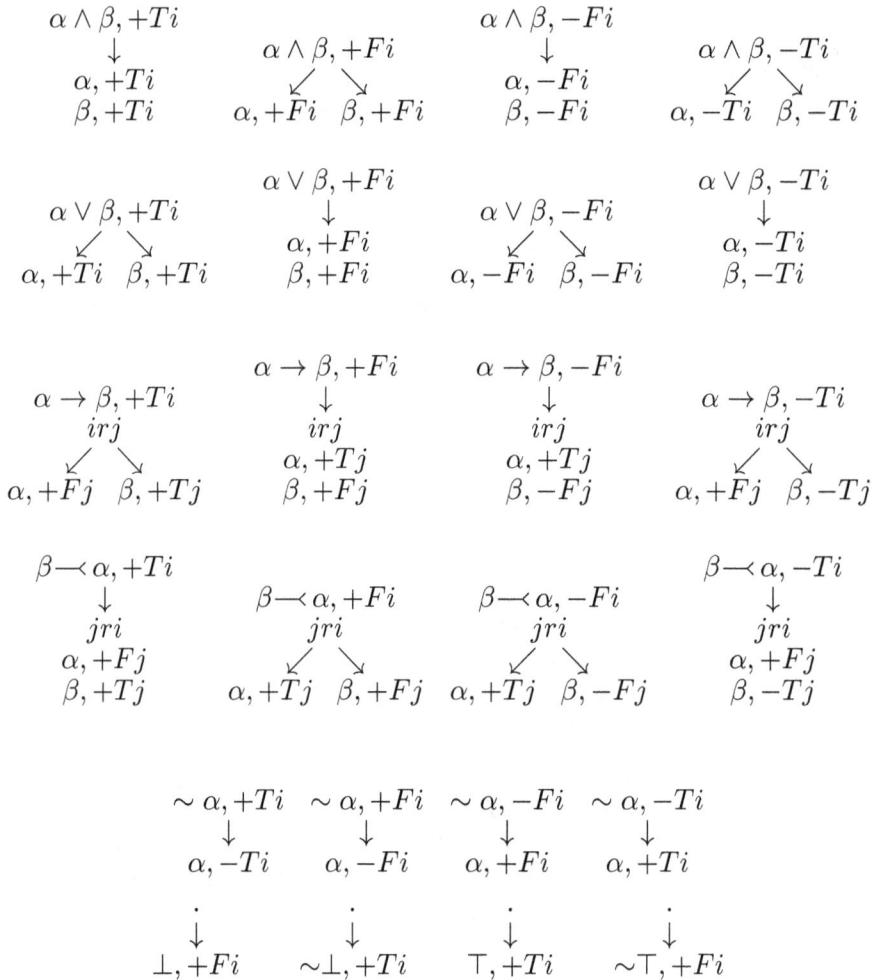

$$\begin{array}{c} \alpha \wedge \beta, +Ti \\ \downarrow \\ \alpha, +Ti \\ \beta, +Ti \end{array} \qquad \begin{array}{c} \alpha \wedge \beta, +Fi \\ \swarrow \quad \searrow \\ \alpha, +Fi \quad \beta, +Fi \end{array} \qquad \begin{array}{c} \alpha \wedge \beta, -Fi \\ \downarrow \\ \alpha, -Fi \\ \beta, -Fi \end{array} \qquad \begin{array}{c} \alpha \wedge \beta, -Ti \\ \swarrow \quad \searrow \\ \alpha, -Ti \quad \beta, -Ti \end{array}$$

$$\begin{array}{c} \alpha \vee \beta, +Ti \\ \swarrow \quad \searrow \\ \alpha, +Ti \quad \beta, +Ti \end{array} \qquad \begin{array}{c} \alpha \vee \beta, +Fi \\ \downarrow \\ \alpha, +Fi \\ \beta, +Fi \end{array} \qquad \begin{array}{c} \alpha \vee \beta, -Fi \\ \swarrow \quad \searrow \\ \alpha, -Fi \quad \beta, -Fi \end{array} \qquad \begin{array}{c} \alpha \vee \beta, -Ti \\ \downarrow \\ \alpha, -Ti \\ \beta, -Ti \end{array}$$

$$\begin{array}{c} \alpha \rightarrow \beta, +Ti \\ irj \\ \swarrow \quad \searrow \\ \alpha, +Fj \quad \beta, +Tj \end{array} \qquad \begin{array}{c} \alpha \rightarrow \beta, +Fi \\ \downarrow \\ irj \\ \alpha, +Tj \\ \beta, +Fj \end{array} \qquad \begin{array}{c} \alpha \rightarrow \beta, -Fi \\ \downarrow \\ irj \\ \alpha, +Tj \\ \beta, -Fj \end{array} \qquad \begin{array}{c} \alpha \rightarrow \beta, -Ti \\ irj \\ \swarrow \quad \searrow \\ \alpha, +Fj \quad \beta, -Tj \end{array}$$

$$\begin{array}{c} \beta \prec \alpha, +Ti \\ \downarrow \\ jri \\ \alpha, +Fj \\ \beta, +Tj \end{array} \qquad \begin{array}{c} \beta \prec \alpha, +Fi \\ jri \\ \swarrow \quad \searrow \\ \alpha, +Tj \quad \beta, +Fj \end{array} \qquad \begin{array}{c} \beta \prec \alpha, -Fi \\ jri \\ \swarrow \quad \searrow \\ \alpha, +Tj \quad \beta, -Fj \end{array} \qquad \begin{array}{c} \beta \prec \alpha, -Ti \\ \downarrow \\ jri \\ \alpha, +Fj \\ \beta, -Tj \end{array}$$

$$\begin{array}{c} \sim\alpha, +Ti \\ \downarrow \\ \alpha, -Ti \end{array} \quad \begin{array}{c} \sim\alpha, +Fi \\ \downarrow \\ \alpha, -Fi \end{array} \quad \begin{array}{c} \sim\alpha, -Fi \\ \downarrow \\ \alpha, +Fi \end{array} \quad \begin{array}{c} \sim\alpha, -Ti \\ \downarrow \\ \alpha, +Ti \end{array}$$

$$\begin{array}{c} \vdots \\ \downarrow \\ \bot, +Fi \end{array} \quad \begin{array}{c} \vdots \\ \downarrow \\ \sim\bot, +Ti \end{array} \quad \begin{array}{c} \vdots \\ \downarrow \\ \top, +Ti \end{array} \quad \begin{array}{c} \vdots \\ \downarrow \\ \sim\top, +Fi \end{array}$$

Structural rules (for capturing the reflexivity and transitivity of the relation $\leq$) and rules for capturing the hereditary condition

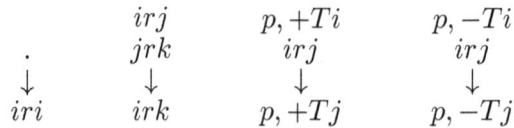

$$\begin{array}{c} \cdot \\ \downarrow \\ iri \end{array} \qquad \begin{array}{c} irj \\ jrk \\ \downarrow \\ irk \end{array} \qquad \begin{array}{c} p, +Ti \\ irj \\ \downarrow \\ p, +Tj \end{array} \qquad \begin{array}{c} p, -Ti \\ irj \\ \downarrow \\ p, -Tj \end{array}$$

The decomposition rules and the hereditary rules in which a statement irj appears above an arrow are applied whenever a node irj occurs on the branch; the decomposition rules in which a statement irj appears above an arrow require the introduction of a new natural number j not already occurring in the tableau. In applications the smallest natural number not already occurring in the tableau is chosen. The structural rule for reflexivity may be used to introduce a node iri when i occurs on the branch. If the application of a rule would result in creating a node that is already on the branch, the rule is not applied. Note that due to the transitivity rule, tableaux may nevertheless be infinite.

We define notions of provability and dual provability.

Definition 5.2. Let $\Delta \cup \{\alpha\}$ be a finite set of $\mathcal{L}_{\mathrm{BCL}}$-formulas. We say that α is provable from Δ ($\Delta \vdash \alpha$) iff there exists a closed and completed tableau for a list of nodes consisting of $\alpha, +F0$ and $\beta, +T0$, for every $\beta \in \Delta$. We say that α is dually provable from Δ ($\Delta \vdash^d \alpha$) iff there exists a closed and completed tableau for a list of nodes consisting of $\alpha, -F0$ and $\beta, -T0$, for every $\beta \in \Delta$.

As an example of a tableau proof we present a proof of Uustalu's [20] counter-example to cut-elimination in the Gentzen-type sequent calculus by Rauszer [24], $p \vdash q \vee (r \to ((p \prec q) \wedge r))$:

$$
\begin{array}{c}
p, +T0 \\
q \vee (r \to ((p \prec q) \wedge r)), +F0 \\
\downarrow \\
q, +F0 \\
r \to ((p \prec q) \wedge r), +F0 \\
\downarrow \\
0r1 \\
r, +T1 \\
(p \prec q) \wedge r, +F1 \\
\swarrow \quad \searrow \\
p \prec q, +F1 \quad\quad r, +F1 \\
\swarrow \quad \searrow \\
p, +F0 \quad q, +T0
\end{array}
$$

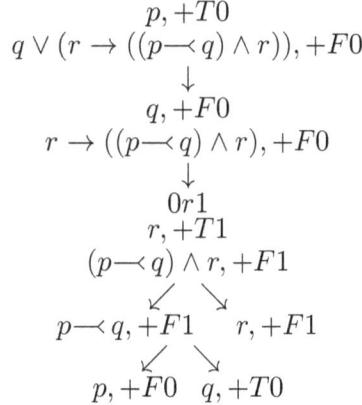

Definition 5.3. Let $\mathcal{M} = \langle M, \leq, \models^+, \models^- \rangle$ be any connexive Kripke model and let br be a tableau branch. The model $\mathcal{M}$ is said to be faithful to br iff there exists a function f from the set of all natural numbers into M such that:

1. for every node $\alpha, +Ti$ on br, $f(i) \models^+ \alpha$;

2. for every node $\alpha, +Fi$ on br, $f(i) \not\models^+ \alpha$;

3. for every node $\alpha, -Ti$ on br, $f(i) \models^- \alpha$;

4. for every node $\alpha, -Fi$ on br, $f(i) \not\models^- \alpha$;

5. for every node jrk on br, $f(j) \leq f(k)$.

The function f is said to show that $\mathcal{M}$ is faithful to branch br.

Lemma 5.4. Let $\mathcal{M}$ be any connexive Kripke model and br be any tableau branch. If $\mathcal{M}$ is faithful to br and a tableau rule is applied to br, then the application produces at least one extension br' of br, such that $\mathcal{M}$ is faithful to br'.

Proof. By induction on the construction of tableaux. We show the cases of the decomposition rules for $\beta \prec \alpha, +Ti$ and $\beta \prec \alpha, +Fi$.

$\beta \prec \alpha, +Ti$: Suppose that the function f shows $\mathcal{M}$ to be faithful to a branch containing $\beta \prec \alpha, +Ti$, so that $f(i) \models^+ \beta \prec \alpha$. Then $\exists y \in M[f(i) \geq y$ and $y \models^+ \beta$ and $y \not\models^+ \alpha]$. An application of the rule for $\beta \prec \alpha, +Ti$ yields new nodes jri, $\alpha, +Fj$, and $\beta, +Tj$. We define the function f' exactly as f except that $f'(j) = y$. Then f' shows that $\mathcal{M}$ is faithful to the the extended branch.

$\beta \prec \alpha, +Fi$: Suppose that the function f shows $\mathcal{M}$ to be faithful to a branch containing $\beta \prec \alpha, +Fi$ and jri, so that $f(j) \leq f(i))$ and $f(i) \not\models^+ \beta \prec \alpha$. Then $\forall y \in M$ $[f(i) \geq y$ implies $(y \not\models^+ \beta$ or $y \models^+ \alpha)]$. Thus $f(j) \not\models^+ \beta$ or $f(j) \models^+ \alpha$ and hence f shows that $\mathcal{M}$ is faithful to at least one expanded branch resulting from the application of the decomposition rule for $\beta \prec \alpha, +Fi$. $\square$

Definition 5.5. Let br be a completed and open tableau branch. Then the structure $\mathcal{M}_{br} = \langle M_{br}, \leq_{br}, \models^+_{br}, \models^-_{br} \rangle$ induced by br is defined as follows:

1. $M_{br} := \{x_j \mid j$ occurs on $br\}$,

2. $x_j \leq_{br} x_k$ iff jrk occurs on br,

3. $x_j \in \models^+_{br} (p)$ iff $p, +Tj$ occurs on br,

4. $x_j \in \models^-_{br} (p)$ iff $p, -Tj$ occurs on br.

Since br is a completed branch, $\leq_{br}$ is reflexive and transitive, and thus $\langle M_{br}, \leq_{br} \rangle$ is a Kripke frame and $\mathcal{M}_{br}$ is a connexive Kripke model. The hereditary condition is satisfied because for any $\star \in \{+, -\}$, if $x_j \models^\star p$ and $x_j \leq_{br} x_k$, then $p, \star Tj$ and jrk occur on br. Since br is completed, the hereditary rule has been applied. Thus, $p, \star Tk$ occurs on br and hence $x_k \models^\star p$. Moreover, since br is an open branch, $x_j \notin \models^+_{br} (p)$ if the node $p, +Fj$ occurs on br, and $x_j \notin \models^-_{br} (p)$ if the node $p, -Fj$ occurs on br.

Lemma 5.6. Suppose that br is a completed and open tableau branch, and let $\mathcal{M}_{br} = \langle M_{br}, \leq_{br}, \models^+_{br}, \models^-_{br} \rangle$ be the model induced by br. Then

- If $\alpha, +Ti$ occurs on br, then $\mathcal{M}_{br}, x_i \models^+ \alpha$

- If $\alpha, +F$ occurs on br, then $\mathcal{M}_{br}, x_i \not\models^+ \alpha$

- If $\alpha, -Ti$ occurs on br, then $\mathcal{M}_{br}, x_i \models^- \alpha$

- If $\alpha, -Fi$ occurs on br, then $\mathcal{M}_{br}, x_i \not\models^- \alpha$.

Proof. By simultaneous induction on the construction of α. If α is a propositional variable, the claim follows by Definition 5.5 and by the fact that we consider an open tableau branch. We present two cases for $\alpha \equiv \beta \to \gamma$.

Let $\beta \to \gamma, +Ti$ occur on br. Since br is completed, if irj occurs on b, then $\beta, +Fj$ or $\gamma, +Tj$ occur on br. Thus, by the induction hypotheses for (5), (2), and (1), if $f(i) \leq_{br} f(j)$, then $f(j) \not\models^+_{br} \alpha$ or $f(j) \models^+_{br} \beta$. But this holds just in case $f(i) \models^+_{br} \beta \to \gamma$.

Let $\beta \to \gamma, +Fi$ occur on br. Since br is completed, the following nodes occur on br for some j: (i) irj, (ii) $\beta, +Tj$, and (iii) $\gamma +Fj$. By the induction hypotheses for (5) and (1), $f(i) \leq_{br} f(j)$, $f(j) \models^+_{br} \alpha$, and $f(j) \not\models^+_{br} \beta$. In other words, $f(i) \not\models^+_{br} \beta \to \gamma$. $\square$

From the previous two lemmas, it follows that the above tableau calculus is sound and complete for BCL with respect to both validity and dual validity.

Theorem 5.7. Let $\Delta \cup \{\alpha\}$ be a finite set of $\mathcal{L}_{\mathrm{BCL}}$-formulas. Then 1. $\Delta \models_{\mathrm{BCL}} \alpha$ iff $\Delta \vdash \alpha$, and 2. $\Delta \models^d_{\mathrm{BCL}} \alpha$ iff $\Delta \vdash^d \alpha$.

Proof. We prove the second claim; the proof for the first claim is analogous.

Soundness: Suppose, by contraposition, that it is not the case that $\Delta \models^d_{\mathrm{BCL}} \alpha$ and let $\Delta = \{\beta_1, \ldots, \beta_n\}$. Then there is a connexive Kripke model $\mathcal{M}$ with a state $x \in M$ such that $x \models^- \beta_1 \ldots x \models^- \beta_n$ but $x \not\models^- \alpha$. The model $\mathcal{M}$ is faithful to the tableau branch consisting of $\beta_1, -T0, \ldots, \beta_n, -T0, \alpha, -F0$. By Lemma 5.4, a completed tableau obtained from that list contains at least one branch to which $\mathcal{M}$ is faithful. Clearly, this branch and hence the tableau must be open.

Completeness: Suppose, by contraposition, that it is not the case that $\Delta \vdash^d \alpha$. Then there is a completed open tableau starting with $\beta_1, -T0, \ldots, \beta_n, -T0, \alpha, -F0$, where $\Delta = \{\beta_1, \ldots, \beta_n\}$. Let br be an open branch of that tableau. By Lemma 5.6, in the model induced by br, the state x_0 reveals that $\Delta \not\models^d_{\mathrm{BCL}} \alpha$. $\square$

Note that sound and complete tableau calculi for the subsystems ICL, DCL, BL, IL, and DL of BCL from Definition 2.2 can be obtained from the tableau calculus for BCL by deleting the decomposition rules for the connectives that are left out in the respective subsystem.

6 Brief outlook

There are many open questions for future research with both a formal and a more philosophical concern. Are there any specific applications of connexive Heyting-Brouwer logic in addition to already known applications of systems of connexive logic? Applications to modelling syllogistic reasoning call for an extension to first order, and so does the discussion about co-implication in Heyting-Brouwer logic as a constructive connective, cf. [9, 40]. Another topic of interest is functional completeness for connexive Heyting-Brouwer logic, either along proof-theoretic lines, see, for example [38], or model-theoretic lines, see, for instance, [14].

References

[1] A. Almukdad and D. Nelson. Constructible falsity and inexact predicates. *Journal of Symbolic Logic*, 49(1):231–233, 1984.

[2] R. B. Angell. A propositional logic with subjunctive conditionals. *Journal of Symbolic Logic*, 27(3):327–343, 1962.

[3] J. Czermak. A remark on Gentzen's calculus of sequents. *Notre Dame Journal of Formal Logic*, 18(3):471–474, 1977.

[4] N. D. Goodman. The logic of contradiction. *Zeitschrift für Mathematische Logik und Grundlagen der Mathematik*, 27(8–10):119–126, 1981.

[5] R. Goré. Dual intuitionistic logic revisited. In R. Dyckhoff, editor, *Automated Reasoning with Analytic Tableaux and Related Methods*, volume 1847 of *Lecture Notes in AI*, pages 252–267, Berlin, 2000. Springer Verlag.

[6] Y. Gurevich. Intuitionistic logic with strong negation. *Studia Logica*, 36(1):49–59, 1977.

[7] N. Kamide and H. Wansing. Connexive modal logic based on positive S4. In J.-Y. Béziau and M. Coniglio, editors, *Logic without Frontiers. Festschrift for Walter Alexandre Carnielli on the Occasion of His 60th Birthday*, pages 389–409. London, College Publications.

[8] N. Kamide and H. Wansing. *Proof Theory of N4-Related Paraconsistent Logics*. London, College Publications, 2015.

[9] E. G. K. López-Escobar. On intuitionistic sentential connectives. I. *Revista Columbiana de Matemáticas*, 19:117–130, 1985.

[10] P. Łukowski. Modal interpretation of Heyting-Brouwer logic. *Bulletin of the Section of Logic*, 25(2):80–83, 1996.

[11] P. Łukowski. A deductive-reductive form of logic: Intuitionistic S4 modalities. *Logic and Logical Philosophy*, 10:79–91, 2002.

[12] S. McCall. Connexive implication. *Journal of Symbolic Logic*, 31(3):415–433, 1966.

[13] S. McCall. A history of connexivity. In D. M. Gabbay *et al.*, editors, *Handbook of the History of Logic*, volume 11, pages 415–449. Elsevier, Amsterdam, 2012.

[14] D. P. McCullough. Logical connectives for intuitionistic propositional logic. *Journal of Symbolic Logic*, 36(1):15–20, 1971.

[15] G. Moisil. Logique modale. *Disquisitiones Mathematicae et Physicae (Bucharest)*, 2:3–98, 1942.

[16] D. Nelson. Constructible falsity. *Journal of Symbolic Logic*, 14(1):16–26, 1949.

[17] S. P. Odintsov. *Constructive Negations and Paraconsistency*. Springer, Dordrecht, 2008.

[18] S. P. Odintsov and H. Wansing. Modal logics with Belnapian truth values. *Journal of Applied Non-Classical Logics*, 20(3):279–301, 2010.

[19] H. Omori and H. Wansing. Web site on connexive logic. 2015.

[20] L. Pinto and T. Uustalu. Relating sequent calculi for bi-intuitionistic propositional logic. In S. van Bakel, S. Berardi, and U. Berger, editors, *Proceedings of the Third International Workshop on Classical Logic and Computation*, volume 47 of *Electronic Proceedings in Theoretical Computer Science*, pages 57–72. 2010.

[21] L. Postniece. *Proof Theory and Proof Search of Bi-Intuitionistic and Tense Logic*. PhD thesis, The Australian National University, Canberra, 2010.

[22] G. Priest. *An Introduction to Non-Classical Logic. From If to Is*. Cambridge University Press, Cambridge, second edition, 2008.

[23] G. Priest. Many-valued modal logics: A simple approach. *Review of Symbolic Logic*, 1(2):190–203, 2008.

[24] C. Rauszer. A formalization of the propositional calculus of H-B logic. *Studia Logica*, 33(1):23–34, 1974.

[25] C. Rauszer. Applications of Kripke models to Heyting-Brouwer logic. *Studia Logica*, 36(1–2):61–71, 1977.

[26] C. Rauszer. An algebraic and Kripke-style approach to a certain extension of intuitionistic logic. *Dissertationes Mathematicae*, 167:1–62, 1980.

[27] W. Rautenberg. *Klassische und nicht-klassische Aussagenlogik*. Vieweg, Braunschweig, 1979.

[28] P. Schroeder-Heister. Schluß und Umkehrschluß. In C. F. Gethmann, editor, *Lebenswelt und Wissenschaft: Ein Beitrag zur Definitionstheorie*, volume 3, pages 1065–1092. Deutsches Jahrbuch Philosophie, 2009.

[29] Y. Shramko. Dual intuitionistic logic and a variety of negations: The logic of scientific research. *Studia Logica*, 80(2):347–367, 2005.

[30] Y. Shramko. A modal translation of dual-intuitionistic logic. *Review of Symbolic Logic*, 2016. to appear.

[31] T. Skolem. Untersuchungen über die Axiome des Klassenkalküls und Über Produktations- und Summationsprobleme, welche gewisse Klassen von Aussagen betreffen. In *Skrifter utgit av Videnskabsselskapet i Kristiania*, volume 3. 1919. reprinted in [32].

[32] T. Skolem. *T. Skolem: Selected Works in Logic*. Universitetforlaget, Oslo, 1970.

[33] G. Takeuti. *Proof Theory*. North-Holland, Amsterdam, 1985.

[34] A. R. Turquette. Review: Gr.C. Moisil, Logique modale. *Journal of Symbolic Logic*, 13(3):162–163, 1948.

[35] I. Urbas. Dual-intuitionistic logic. *Notre Dame Journal of Formal Logic*, 37(3):440–451, 1996.

[36] N. N. Vorob'ev. A constructive propositional calculus with strong negation. Doklady Akademii Nauk SSR, 85:465–468, 1952. (in Russian).

[37] H. Wansing. Falsification, natural deduction and bi-intuitionistic logic. *Journal of Logic and Computation*, 26(2016):425–450. First published online July 17, 2013.

[38] H. Wansing. Functional completeness for subsystems of intuitionistic propositional logic. *Journal of Philosophical Logic*, 22(3):303–321, 1993.

[39] H. Wansing. Connexive modal logic. 5:367–383, 2005.

[40] H. Wansing. Constructive negation, implication, and co-implication. *Journal of Applied Non-Classical Logics*, 18(2–3):341–364, 2008.

[41] H. Wansing. Connexive logic. In E. N. Zalta, editor, *The Stanford Encyclopedia of Philosophy*. Fall 2014 edition, 2014. `http://plato.stanford.edu/archives/fall2010/entries/logic-connexive/`.

[42] H. Wansing. Natural deduction for bi-connexive logic and a two-sorted typed λ-calculus. *IFCoLog Journal of Logic and their Applications*, 2016. this issue.

[43] H. Wansing. On split negation, strong negation, information, falsification, and verification. In K. Bimbó, editor, *J. Michael Dunn on Information Based Logics*, volume 8 of *Outstanding Contributions to Logic*, pages 161–189. Springer, Dordrecht, 2016.

       Received October 2016

A Simple Connexive Extension of the Basic Relevant Logic BD

Hitoshi Omori
Department of Philosophy
Kyoto University, Japan
`hitoshiomori@gmail.com`

Abstract

Motivated by an open problem formulated by Graham Priest and Richard Sylvan related to the basic relevant logic **BD**, the present note offers a partial solution to the problem by making use of an idea suggested by Heinrich Wansing in the context of connexive logic. The note also presents two other non-connexive options that can be regarded as partial solutions to the problem.

1 Introduction

The name 'Connexive logic' suggests that connexive logic shares a certain motivation with relevant logic.[1] And some attempts are known in the literature by relevantists such as Richard Routley, Chris Mortensen and Ross Brady, at realizing the connexive theses in relevant logic. The present paper goes in the same direction using a different approach. The main motivation behind the paper involves a problem formulated by Graham Priest and Richard Sylvan in [8], and considered further by Greg Restall in [9, 10]. In brief, the problem is to find a proof theory for extensions of the basic relevant logic **BD** in which the negation is interpreted in terms of a four-valued semantics (i.e. the so-called American plan). The difficulty lies in finding the appropriate axioms and/or rules of inference, to capture the corresponding falsity condition for the conditional. Priest and Sylvan suggested two falsity conditions for the conditional, but the corresponding axioms and/or rules of inference remain unknown. The aim of this note is to show that for a certain falsity condition, inspired

---

The author is a Postdoctoral Research Fellow of the Japan Society for the Promotion of Science (JSPS). I would like to thank Graham Priest for directing my attention to the problem discussed in this note, and also Heinrich Wansing for his encouragement.

[1]For an up-to-date survey on connexive logic, see [13].

by the work [12] on connexive logic by Heinrich Wansing, it is possible to find the corresponding proof theory. This note also presents two other non-connexive falsity conditions for which the corresponding proof theories are available.

2 Revisiting the basics for the basic relevant logic BD

In this section, some of the main notions and results from [8] are reviewed which will be used in the main observation of the note.

Definition 1. The language $\mathcal{L}$ consists of a finite set $\{\sim, \wedge, \vee, \rightarrow\}$ of propositional connectives and a countable set Prop of propositional parameters which we denote by p, q, etc. Furthermore, we denote by Form the set of formulas defined as usual in $\mathcal{L}$. We denote a formula of $\mathcal{L}$ by A, B, C, etc. and a set of formulas of $\mathcal{L}$ by Γ, Δ, Σ, etc.

2.1 Proof theory

Definition 2. The axioms of **BD** are as follows:

(A1) $\qquad A \rightarrow A$

(A2) $A \rightarrow (A \vee B)$ $\quad B \rightarrow (A \vee B)$

(A3) $(A \wedge B) \rightarrow A$ $\quad (A \wedge B) \rightarrow B$

(A4) $(A \wedge (B \vee C)) \rightarrow ((A \wedge B) \vee C)$

(A5) $((A \rightarrow B) \wedge (A \rightarrow C)) \rightarrow (A \rightarrow (B \wedge C))$

(A6) $((A \rightarrow C) \wedge (B \rightarrow C)) \rightarrow ((A \vee B) \rightarrow C)$

(A7) $\sim (A \wedge B) \leftrightarrow (\sim A \vee \sim B)$

(A8) $\sim (A \vee B) \leftrightarrow (\sim A \wedge \sim B)$

(A9) $\sim\sim A \leftrightarrow A$

If

$$\frac{A_1 \ldots A_n}{B}$$

is a rule scheme, then we define its disjunctive form to be the scheme

$$\frac{(C \vee A_1) \ldots (C \vee A_n)}{C \vee B}.$$

The rules for **BD** are the following plus their disjunctive forms:

(R1) $\dfrac{A \quad A \rightarrow B}{B}$ (R2) $\dfrac{A \quad B}{A \wedge B}$ (R3) $\dfrac{A \rightarrow B \quad C \rightarrow D}{(B \rightarrow C) \rightarrow (A \rightarrow D)}.$

Finally, if Σ is a set of formulas and A is a formula, then $\Sigma \vdash A$ is defined in the standard classical fashion.

Remark 3. It deserves noting that the following rules, known as Prefixing, Suffixing and Transitivity respectively, are derivable in **BD** in view of (R3), (R1) and (A1):

$$\frac{C{\to}D}{(A{\to}C){\to}(A{\to}D)} \qquad \frac{A{\to}B}{(B{\to}C){\to}(A{\to}C)} \qquad \frac{A{\to}B \quad B{\to}C}{A{\to}C}.$$

2.2 Semantics

Definition 4. An interpretation for the language is a four-tuple $\langle g, W, R, I \rangle$, where

- W is a set (of worlds);
- $g \in W$ (the base world);
- R is a ternary relation on W;
- I assigns to each pair consisting of a world, w, and propositional parameter, p, a truth value $I(w, p) \in \{\{1\}, \{1, 0\}, \emptyset, \{0\}\}$.

Truth values at worlds are then assigned to all formulas by the following conditions:

$1 \in I(w, {\sim}A)$ iff $0 \in I(w, A)$

$0 \in I(w, {\sim}A)$ iff $1 \in I(w, A)$

$1 \in I(w, A{\wedge}B)$ iff $1 \in I(w, A)$ and $1 \in I(w, B)$

$0 \in I(w, A{\wedge}B)$ iff $0 \in I(w, A)$ or $0 \in I(w, B)$

$1 \in I(w, A{\vee}B)$ iff $1 \in I(w, A)$ or $1 \in I(w, B)$

$0 \in I(w, A{\vee}B)$ iff $0 \in I(w, A)$ and $0 \in I(w, B)$

$1 \in I(w, A{\to}B)$ iff for all $x, y \in W$: if $Rwxy$ and $1 \in I(x, A)$ then $1 \in I(y, B)$

Note here that the falsity of a conditional is arbitrary. Furthermore, we assume that $Rgxy$ iff $x = y$.

Finally, semantic consequence is now defined in terms of truth preservation at g:

$$\Sigma \models A \text{ iff for all } \langle g, W, R, I \rangle, 1 \in I(g, A) \text{ if } 1 \in I(g, B) \text{ for all } B \in \Sigma.$$

Remark 5. Truth and falsity conditions for the $\to$-free fragment are exactly as in the four-valued logic of Belnap and Dunn, also known as First Degree Entailment.

2.3 Soundness and Completeness

For the sake of making the paper self-contained as much as possible, the soundness and completeness proofs are briefly reviewed without the details. The results are entirely due to [8] in which details of the proofs are spelled out.

Theorem 1 (Priest & Sylvan). *For any $\Gamma \cup \{A\} \subseteq$ Form, if $\Gamma \vdash A$ then $\Gamma \models A$.*

Proof. The proof is by a simple induction over the length of proofs, as usual. □

Definition 6. We introduce the following notions.

1. If Π is a set of sentences, let $\Pi_\rightarrow$ be the set of all members of Π of the form $A{\rightarrow}B$.
2. $\Sigma \vdash_\pi A$ iff $\Sigma \cup \Pi_\rightarrow \vdash A$.
3. Σ is a Π-*theory* iff:

   (a) if $A, B{\in}\Sigma$ then $A{\wedge}B{\in}\Sigma$

   (b) if $\vdash_\pi A{\rightarrow}B$ then (if $A{\in}\Sigma$ then $B{\in}\Sigma$).
4. Σ is *prime* iff (if $A{\vee}B{\in}\Sigma$ then $A{\in}\Sigma$ or $B{\in}\Sigma$).
5. If X is any set of sets of formulas the ternary relation R on X is defined thus:

$$R\Sigma\Gamma\Delta \text{ iff } (\text{if } A{\rightarrow}B{\in}\Sigma \text{ then } (\text{if } A{\in}\Gamma \text{ then } B{\in}\Delta)).$$

6. $\Sigma \vdash_\pi \Delta$ iff for some $D_1,\ldots,D_n{\in}\Delta, \Sigma \vdash_\pi D_1{\vee}\ldots{\vee}D_n$.
7. $\vdash_\pi \Sigma{\rightarrow}\Delta$ iff for some $C_1,\ldots,C_n{\in}\Sigma$ and $D_1,\ldots,D_m{\in}\Delta$:

$$\vdash_\pi C_1{\wedge}\ldots{\wedge}C_n{\rightarrow}D_1{\vee}\ldots{\vee}D_n.$$

8. Σ is Π-*deductively closed* iff (if $\Sigma \vdash_\pi A$ then $A{\in}\Sigma$).
9. $\langle\Sigma,\Delta\rangle$ is a Π-*partition* iff:

   (a) $\Sigma \cup \Delta = \mathsf{Form}$

   (b) $\nvdash_\pi \Sigma{\rightarrow}\Delta$

In all the above, if Π is $\emptyset$, then the prefix 'Π-' and the subscript π will simply be omitted.

With these notions in mind, some lemmas are reviewed without their proofs. The first group concerns extensions of sets with various properties.

Lemma 1. *If $\langle\Sigma,\Delta\rangle$ is a Π-partition then Σ is a prime Π-theory.*

Lemma 2. *If $\nvdash_\pi \Sigma{\rightarrow}\Delta$ then there are $\Sigma' \supseteq \Sigma$ and $\Delta' \supseteq \Delta$ such that $\langle\Sigma',\Delta'\rangle$ is a Π-partition.*

Corollary 1. *Let Σ be a Π-theory, Δ be closed under disjunction, and $\Sigma \cap \Delta = \emptyset$. Then there is $\Sigma' \supseteq \Sigma$ such that $\Sigma' \cap \Delta = \emptyset$ and Σ' is a prime Π-theory.*

Lemma 3. *If $\Sigma \nvdash \Delta$ then there are $\Sigma' \supseteq \Sigma$ and $\Delta' \supseteq \Delta$ such that $\langle\Sigma',\Delta'\rangle$ is a partition, and Σ' is deductively closed.*

Corollary 2. *If $\Sigma \nvdash A$ then there is $\Pi \supseteq \Sigma$ such that $A{\notin}\Pi$, Π is a prime Π-theory and Π is Π-deductively closed.*

The second group of lemmas establishes that there are certain theories with properties that are crucial in the proof of the main theorem as far as the conditional is concerned.

Lemma 4. *If Π is a prime Π-theory, is Π-deductively closed and $A{\rightarrow}B{\notin}\Pi$, then there is a prime Π-theory Γ, such that $A{\in}\Gamma$ and $B{\notin}\Gamma$.*

Lemma 5. *If Σ, Γ, Δ are Π-theories, $R\Sigma\Gamma\Delta$ and $A{\notin}\Delta$, then there are prime Π-theories, Γ', Δ', such that $\Gamma' \supseteq \Gamma$, $A{\notin}\Delta'$ and $R\Sigma\Gamma'\Delta'$.*

Lemma 6. *Let Σ be a prime Π-theory and $A{\rightarrow}B{\notin}\Sigma$. Then there are prime Π-theories, Γ', Δ' such that $R\Sigma\Gamma'\Delta'$, $A{\in}\Gamma'$, $B{\notin}\Delta'$.*

By making use of these lemmas, we are now ready to prove the following completeness theorem.

Theorem 2 (Priest & Sylvan). *For any $\Gamma \cup \{A\} \subseteq$ Form, if $\Gamma \models A$ then $\Gamma \vdash A$.*

Proof. We prove the contrapositive. Suppose that $\Gamma \nvdash A$. Then, by Corollary 2, there is a $\Pi \supseteq \Gamma$ such that Π is a prime theory and $A{\notin}\Pi$. Define the interpretation $\mathfrak{A} = \langle \Pi, X, R, I \rangle$, where $X = \{\Delta : \Delta$ is a prime Π-theory$\}$, R as in Definition 6 and I is defined thus. For every state Σ, $p \in$ Prop and $C, D \in$ Form:

$$1{\in}I(\Sigma, p) \text{ iff } p{\in}\Sigma, \quad 0{\in}I(\Sigma, p) \text{ iff } {\sim}p{\in}\Sigma, \quad 0{\in}I(\Sigma, C \rightarrow D) \text{ iff } {\sim}(C \rightarrow D){\in}\Sigma.$$

We show that the following condition holds for any arbitrary formula, B:

$(*)$ $\qquad\qquad 1{\in}I(\Sigma, B)$ iff $B{\in}\Sigma$ and $0{\in}I(\Sigma, B)$ iff ${\sim}B{\in}\Sigma$

It then follows that $\mathfrak{A}$ is a counter-model for the inference, and hence that $\Gamma \nvDash A$. The proof of $(*)$ is by induction on the complexity of B.
For negation: We begin with the positive clause.

$$\begin{aligned} 1{\in}I(\Sigma, {\sim}C) &\text{ iff } 0{\in}I(\Sigma, C) \\ &\text{ iff } {\sim}C{\in}\Sigma \qquad\qquad \text{IH} \end{aligned}$$

Again, the negative clause is also straightforward.

$$\begin{aligned} 0{\in}I(\Sigma, {\sim}C) &\text{ iff } 1{\in}I(\Sigma, C) \\ &\text{ iff } C{\in}\Sigma \qquad\qquad \text{IH} \\ &\text{ iff } {\sim}{\sim}C{\in}\Sigma \qquad\qquad \text{(A9)} \end{aligned}$$

For disjunction: We begin with the positive clause.

$1{\in}I(\Sigma, C{\vee}D)$ iff $1{\in}I(\Sigma, C)$ or $1{\in}I(\Sigma, D)$

    iff $C{\in}\Sigma$ or $D{\in}\Sigma$ IH

    iff $C{\vee}D{\in}\Sigma$ Σ is a prime theory

The negative clause is also straightforward.

$0{\in}I(\Sigma, C{\vee}D)$ iff $0{\in}I(\Sigma, C)$ and $0{\in}I(\Sigma, D)$

    iff ${\sim}C{\in}\Sigma$ and ${\sim}D{\in}\Sigma$ IH

    iff ${\sim}C{\wedge}{\sim}D{\in}\Sigma$ Σ is a theory

    iff ${\sim}(C{\vee}D){\in}\Sigma$ (A8)

For conjunction: We begin with the positive clause.

$1{\in}I(\Sigma, C{\wedge}D)$ iff $1{\in}I(\Sigma, C)$ and $1{\in}I(\Sigma, D)$

    iff $C{\in}\Sigma$ and $D{\in}\Sigma$ IH

    iff $C{\wedge}D{\in}\Sigma$ Σ is a theory

Again, the negative clause is also straightforward.

$0{\in}I(\Sigma, C{\wedge}D)$ iff $0{\in}I(\Sigma, C)$ or $0{\in}I(\Sigma, D)$

    iff ${\sim}C{\in}\Sigma$ or ${\sim}D{\in}\Sigma$ IH

    iff ${\sim}C{\vee}{\sim}D{\in}\Sigma$ Σ is a prime theory

    iff ${\sim}(C{\wedge}D){\in}\Sigma$ (A7)

For the conditional: We split the case depending on $\Sigma = \Pi$ or not. First, if $\Sigma = \Pi$, then we have the following:

$1{\in}I(\Pi, C{\to}D)$ iff $\forall\Gamma{\in}X$(if $1{\in}I(\Gamma, C)$ then $1{\in}I(\Gamma, D)$)

    iff $\forall\Gamma{\in}X$(if $C{\in}\Gamma$ then $D{\in}\Gamma$)   IH

    iff $C{\to}D{\in}\Pi$

For the last equivalence, the top-to-bottom direction holds by Lemma 4 and the bottom-to-top direction holds by the fact that Γ is a Π-theory.

Second, if $\Sigma \neq \Pi$, then we have the following:

$1{\in}I(\Sigma, C{\to}D)$ iff for all Γ, Δ s.t. $R\Sigma\Gamma\Delta$, if $1{\in}I(\Gamma, C)$ then $1{\in}I(\Delta, D)$

    iff for all Γ, Δ s.t. $R\Sigma\Gamma\Delta$, if $C{\in}\Gamma$ then $D{\in}\Delta$ IH iff $C{\to}D{\in}\Sigma$

For the last equivalence, the top-to-bottom direction holds by Lemma 6 and the bottom-to-top direction holds by the definition of R (cf. Definition 6). Thus, we obtain the desired result. $\square$

3 Main observations

First, three falsity conditions for the conditional are introduced.

Definition 7. Consider the following falsity conditions for the conditional.

- $0{\in}I(w, A{\to}B)$ iff for some $x, y{\in}W$: $Rwxy$ and $1{\in}I(x, A)$ and $1{\notin}I(y, B)$.
- $0{\in}I(w, A{\to}B)$ iff $1{\in}I(w, A)$ and $0{\in}I(w, B)$.
- $0{\in}I(w, A{\to}B)$ iff for all $x, y{\in}W$: if $Rwxy$ and $1{\in}I(x, A)$ then $0{\in}I(y, B)$.

We define $\models_1, \models_2$ and $\models_3$ as semantic consequence relations obtained by adding the above conditions to the semantics for **BD** respectively.

Remark 8. The first condition is a variant of the following condition suggested by Priest and Sylvan:

$$0{\in}I(w, A{\to}B) \text{ iff for some } x, y{\in}W: Rwxy, 1{\in}I(x, A) \text{ and } 0{\in}I(y, B).$$

Moreover, the second condition is exactly the condition we find in the study of constructive falsity by David Nelson (cf. [4]), followed by further systematic studies by Norihiro Kamide, Sergei Odintsov and Heinrich Wansing among others (cf. [5, 2]). Finally, the third condition is a natural variant of the connexive conditional studied by Wansing in [12].

Second, three extensions of **BD** are introduced.

Definition 9. Consider the following formulas.

(AxS1) $C{\to}((A{\to}B){\vee}{\sim}(A{\to}B))$ (AxN) ${\sim}(A{\to}B){\leftrightarrow}(A{\wedge}{\sim}B)$

(AxS2) $((A{\to}B){\wedge}{\sim}(A{\to}B)){\to}C$ (AxW) ${\sim}(A{\to}B){\leftrightarrow}(A{\to}{\sim}B)$

Then we introduce the following three systems.

- **BDS**: **BD** with (AxS1) and (AxS2) as additional axioms.
- **BDN**: **BD** with (AxN) as an additional axiom.
- **BDW**: **BD** with (AxW) as an additional axiom.

We define proof theoretic consequence relations $\vdash_S, \vdash_N$ and $\vdash_W$ for **BDS, BDN** and **BDW** respectively, as usual.

Remark 10. BDS is named after Antonio Sette who devised a system of paraconsistent logic known as $\mathbf{P}^1$ in which *every* complex formula is explosive (cf. [11]). **BDN** is named after Nelson who introduced the expansion of intuitionistic logic enriched by the so-called strong negation since the axiom (AxN) is the one used in his system. Finally, **BDW** is named after Wansing who introduced and studied the axiom (AxW) in [12].

We then obtain the following results *without* any additional lemma.

Theorem 3. *For any* $\Gamma \cup \{A\} \subseteq$ Form, *we have the following results:*

- $\Gamma \models_1 A$ *iff* $\Gamma \vdash_S A$.
- $\Gamma \models_2 A$ *iff* $\Gamma \vdash_N A$.
- $\Gamma \models_3 A$ *iff* $\Gamma \vdash_W A$.

Proof. The soundness part is relatively straightforward. For the completeness part, the proof runs exactly as in the case for **BD**. For both directions, the only thing to be checked is the negative clause for conditionals.

For **BDS**: for the soundness, observe the following for any $w \in W$:

$$1 \in I(w, (A \to B) \vee \sim (A \to B))$$
$$\text{iff } 1 \in I(w, A \to B) \text{ or } 0 \in I(w, A \to B)$$
$$\text{iff (for all } x, y \in W: \text{ if } Rwxy \text{ and } 1 \in I(x, A) \text{ then } 1 \in I(y, B)) \text{ or}$$
$$\text{(for some } x_0, y_0 \in W: Rwx_0y_0 \text{ and } 1 \in I(x_0, A) \text{ and } 1 \notin I(y_0, B))$$

Moreover, for any $w \in W$, we have the following:

$$1 \in I(w, (A \to B) \wedge \sim (A \to B))$$
$$\text{iff } 1 \in I(w, A \to B) \text{ and } 0 \in I(w, A \to B)$$
$$\text{iff (for all } x, y \in W: \text{ if } Rwxy \text{ and } 1 \in I(x, A) \text{ then } 1 \in I(y, B)) \text{ and}$$
$$\text{(for some } x_0, y_0 \in W: Rwx_0y_0 \text{ and } 1 \in I(x_0, A) \text{ and } 1 \notin I(y_0, B))$$

For the completeness, we split the case as in the positive clause. First, if $\Sigma = \Pi$, then we have the following:

$$0 \in I(\Pi, C \to D) \text{ iff } \exists \Gamma \in X(1 \in I(\Gamma, C) \text{ and } 1 \notin I(\Gamma, D))$$
$$\text{iff } \exists \Gamma \in X(C \in \Gamma \text{ and } D \notin \Gamma) \qquad \text{IH}$$

$$\text{iff } C{\to}D{\notin}\Pi \qquad\qquad (*)$$
$$\text{iff } {\sim}(C{\to}D){\in}\Pi$$

Note that $(*)$ is the argument given in the case for the positive clause for the conditional. Moreover, for the last equivalence, the top-to-bottom direction holds in view of (AxS1) and the bottom-to-top direction holds in view of (AxS2).

Second, if $\Sigma \neq \Pi$, then we have the following:

$$0{\in}I(\Sigma, C{\to}D) \text{ iff } \exists\Gamma, \Delta{\in}X(R\Sigma\Gamma\Delta \text{ and } 1{\in}I(\Gamma, C) \text{ and } 1{\notin}I(\Delta, D))$$
$$\text{iff } \exists\Gamma, \Delta{\in}X(R\Sigma\Gamma\Delta \text{ and } C{\in}\Gamma \text{ and } D{\notin}\Delta) \qquad \text{IH}$$
$$\text{iff } C{\to}D{\notin}\Sigma \qquad\qquad (\star)$$
$$\text{iff } {\sim}(C{\to}D){\in}\Sigma \qquad\qquad (\dagger)$$

Note that $(\star)$ is again the argument given in the case for the positive clause for the conditional. As for $(\dagger)$, we make use of (AxS1) and (AxS2), as expected.

For **BDN**: for the soundness, observe that the following holds for any $w{\in}W$:

$$1{\in}I(w, {\sim}(A{\to}B)) \text{ iff } 0{\in}I(w, A{\to}B)$$
$$\text{iff } 1{\in}I(w, A) \text{ and } 0{\in}I(w, B)$$
$$\text{iff } 1{\in}I(w, A) \text{ and } 1{\in}I(w, {\sim}B)$$
$$\text{iff } 1{\in}I(w, A{\wedge}{\sim}B)$$

For the completeness, the following shows that we have the desired result.

$$0{\in}I(\Sigma, C{\to}D) \text{ iff } 1{\in}I(\Sigma, C) \text{ and } 0{\in}I(\Sigma, D)$$
$$\text{iff } C{\in}\Sigma \text{ and } {\sim}D{\in}\Sigma \qquad\qquad \text{IH}$$
$$\text{iff } C{\wedge}{\sim}D{\in}\Sigma \qquad\qquad \Sigma \text{ is a theory}$$
$$\text{iff } {\sim}(C{\to}D){\in}\Sigma \qquad\qquad (\text{AxN})$$

For **BDW**: for the soundness, observe that the following holds for any $w{\in}W$:

$$1{\in}I(w, {\sim}(A{\to}B)) \text{ iff } 0{\in}I(w, A{\to}B)$$
$$\text{iff for all } x, y{\in}W\text{: if } Rwxy \text{ and } 1{\in}I(x, A) \text{ then } 0{\in}I(y, B)$$
$$\text{iff for all } x, y{\in}W\text{: if } Rwxy \text{ and } 1{\in}I(x, A) \text{ then } 1{\in}I(y, {\sim}B)$$
$$\text{iff } 1{\in}I(w, A{\to}{\sim}B)$$

For the completeness, we split the case as in the positive condition. First, if $\Sigma = \Pi$, then we have the following:

$$0{\in}I(\Pi, C{\to}D) \text{ iff } \forall\Gamma{\in}X(\text{if } 1{\in}I(\Gamma, C) \text{ then } 0{\in}I(\Gamma, D))$$

$$\text{iff } \forall\Gamma\in X(\text{if } C\in\Gamma \text{ then } \sim D\in\Gamma) \qquad \text{IH}$$
$$\text{iff } C\to\sim D\in\Pi \qquad (*)$$
$$\text{iff } \sim(C\to D)\in\Pi \qquad (\text{AxW})$$

Second, if $\Sigma \neq \Pi$, then we have the following:

$$0\in I(\Sigma, C\to D)$$
$$\text{iff } \forall\Gamma, \Delta\in X(\text{if } R\Sigma\Gamma\Delta \text{ and } 1\in I(\Gamma, C) \text{ then } 0\in I(\Delta, D))$$
$$\text{iff } \forall\Gamma, \Delta\in X(\text{if } R\Sigma\Gamma\Delta \text{ and } C\in\Gamma \text{ then } \sim D\in\Delta) \qquad \text{IH}$$
$$\text{iff } C\to\sim D\in\Sigma \qquad (*)$$
$$\text{iff } \sim(C\to D)\in\Sigma \qquad (\text{AxW})$$

Note here again that $(*)$ is the argument given in the case of the positive clause for the conditional. This completes the proof. $\qquad\square$

4 Conclusion: reflections and the original open problem

4.1 Some reflections on BDW

Compared to the previous attempts in the literature at adding some connexive flavor to relevant logics, such as [3] by Mortensen and [1] by Brady, the system **BDW** is obtained in an extremely simple manner: nothing is required on top of the usual semantic framework for relevant logics. This simplicity is made possible by choosing the American plan rather than the Australian plan in interpreting the negation. Indeed, if we take the Australian plan and interpret the negation in terms of the Routley star, then we need to take additional care about the interaction between the Routley star and the ternary relation to realize the connexive theses, and this brings in further complications as we can find in [3, 1]. However, if we follow the American plan, then we have a simple account of negation which flip-flops truth and falsity, and so the connexive theses will be closely related to the problem of the falsity condition of the conditional. And as observed in the previous section, we may import the idea of Wansing who developed a connexive variant of Nelson logics by introducing a simple falsity condition of the conditional.

Note here that with the Australian plan, we are forced to accept the rule of contraposition, namely the following rule:

$$(\text{Contra}) \qquad\qquad \frac{A\to B}{\sim B\to\sim A}$$

This rule plays a substantial role in stronger systems. For example, Mortensen proves that the relevant logic **R** with Aristotle's thesis, namely $\sim(A{\to}\sim A)$, is trivial (cf. [3, p.109, Theorem 2]). This might make us question if the system **BDW** is *really* non-trivial. The answer is positive, i.e. that **BDW** is indeed non-trivial, since **BDW** is a subsystem of **C** of Wansing, introduced in [12], and **C** is non-trivial. As a related remark, note that **BDW** is contradictory. Indeed, both $(A\wedge\sim A){\to}A$ and its negation are provable in **BDW**.[2]

Finally, one might question the philosophical adequacy of the condition. The details need to be left for another occasion, but if one is in favor of the ternary relation, then **BDW** seems to be quite reasonable.[3]

4.2 Some reflections on BDS and BDN

The system **BDS** has a falsity condition for the conditional which is quite similar to the one suggested by Priest and Sylvan. Moreover, it is proved in Theorem 3 that it has a smooth proof theory. However, seen from the proof-theoretic perspective, (AxS2) destroys the relevance and thus probably is not a reasonable option for relevantists.

Finally, the system **BDN** has an extremely simple falsity condition, as simple as those of the extensional connectives. Note that the additional axiom (AxN) causes some troubles when the negation is interpreted in terms of the Routley star. Indeed, (AxN) together with (Contra) proves *ex contradictione quodlibet*.[4] This is, of course, a bad news for relevantists. However, in our case, this will not be the case since **BDN** is a subsystem of **N4**, one of the Nelson logics, and **N4** is paraconsistent. But then are there any problems with the concerned extension of **BD**? Not that I see it at the moment. I will leave this question for interested readers.

4.3 The original open problem

The original problem formulated by Priest and Sylvan remains open. The problem is to find appropriate axioms and/or rules of inference, to capture the corresponding falsity conditions for the conditional in addition to **BD**:

- $0{\in}I(w, A{\to}B)$ iff for some $x, y{\in}W$: $Rwxy$, $0{\notin}I(x, A)$ and $0{\in}I(y, B)$.

---

[2]A similar result is reported already in [3, p.108, Theorem 1] and [13], though the formula here is slightly simplified.

[3]For an interesting argument for the ternary relation of the relevant logic $\mathbf{B}^{+}$, see [7].

[4]This can be proved as follows. By (AxN), we have $\sim(A \to B) \to (A \wedge \sim B)$, and thus $\sim(A \to B) \to A$ in view of (A4) and Transitivity (cf. Remark 3). Now by applying (Contra), we have $\sim A \to \sim\sim(A \to B)$. Finally, by (A9) and Transitivity, we obtain $\sim A \to (A \to B)$, as desired.

- $0 \in I(w, A \to B)$ iff for some $x, y \in W$: $Rwxy$, $1 \in I(x, A)$ and $0 \in I(y, B)$.

Despite the philosophical worry of Priest about the simplified semantics in general, recently addressed in [6], the problem is interesting from a purely technical perspective.

References

[1] R. Brady. A Routley-Meyer affixing style semantics for logics containing Aristotle's thesis. *Studia Logica*, 48(2):235–241, 1989.

[2] N. Kamide and H. Wansing. *Proof Theory of N4-related Paraconsistent Logics*. Studies in Logic, Vol. 54. College Publications, London, 2015.

[3] C. Mortensen. Aristotle's thesis in consistent and inconsistent logics. *Studia Logica*, 43(1/2):107–116, 1984.

[4] D. Nelson. Constructible falsity. *Journal of Symbolic Logic*, 14(1):16–26, 1949.

[5] S. P. Odintsov. *Constructive Negations and Paraconsistency*. Springer-Verlag, Dordrecht, 2008.

[6] G. Priest. Fusion and confusion. *Topoi*, 34(1):55–61, 2015.

[7] G. Priest. Is the ternary R depraved? In C. Caret and O. Hjortland, editors, *Foundations of Logical Consequence*, pages 121–135. Oxford University Press, 2015.

[8] G. Priest and R. Sylvan. Simplified semantics for basic relevant logic. *Journal of Philosophical Logic*, 21(2):217–232, 1992.

[9] G. Restall. Simplified semantics for relevant logics (and some of their rivals). *Journal of Philosophical Logic*, 22(5):481–511, 1993.

[10] G. Restall. Four-valued semantics for relevant logics (and some of their rivals). *Journal of Philosophical Logic*, 24(2):139–160, 1995.

[11] A. Sette. On the propositional calculus P^1. *Mathematica Japonicae*, 16:173–180, 1973.

[12] H. Wansing. Connexive modal logic. In R. Schmidt, I. Pratt-Hartmann, M. Reynolds, and H. Wansing, editors, *Advances in Modal Logic. Volume 5*, pages 367–383. King's College Publications, 2005.

[13] H. Wansing. Connexive logic. In E. N. Zalta, editor, *The Stanford Encyclopedia of Philosophy*. Fall 2014 edition, 2014. Available at http://plato.stanford.edu/archives/fall2014/entries/logic-connexive/.

 Received January 2016

Natural Deduction for Two Connexive Logics

Nissim Francez
Department of Computer Science
The Technion-IIT, Israel
`francez@cs.technion.ac.il`

Abstract

I propose two natural-deduction proof-systems $\mathcal{N}^{\neg r}$ and $\mathcal{N}^{\neg l}$, for non-classical interactions of a certain kind between negation and implication, that can be seen as variants of connexive logics. These interactions are inspired by a certain use of negation and implication in natural language. I propose the natural-deduction systems as meaning-conferring proof-systems, not appealing to any many-valued model theory as a semantics. The model-theory is used mainly as an auxiliary tool for establishing non-derivability, for example of some classical formal theorems (or, more generally, classical derivability claims) that are not provable (not derivable) in $\mathcal{N}^{\neg r}$ and $\mathcal{N}^{\neg l}$. The relation between implication and negation in the system $\mathcal{N}^{\neg r}$ is similar to the one by Cantwell and one by Cooper, the former unaware of the latter. The system $\mathcal{N}^{\neg l}$ seems to be new.

1 Introduction

In this paper, I propose two natural-deduction (ND) proof-systems $\mathcal{N}^{\neg r}$ and $\mathcal{N}^{\neg l}$, for non-classical interactions of a certain kind between negation and implication,

A talk based on this paper was presented in the workshop *Inferences and Proof*, University of Aix-Marseille, Marseille, May 31 - June 1, 2016. I thank Heinrich Wansing, Hitoshi Omori and an anonymous referee for critical comments that have improved the presentation of the ideas in this paper.

that can be seen as variants of *connexive logics*. These interactions are inspired by a certain use of negation and implication in natural language. I propose the natural-deduction systems $\mathcal{N}^{\neg r}$ and $\mathcal{N}^{\neg l}$ as meaning-conferring proof-systems (see Section 4), not appealing to any many-valued model theory as a semantics. The model-theory (in Section 5) is used mainly as an auxiliary tool for establishing non-derivability, for example of some classical formal theorems (or, more generally, classical derivability claims) that are not provable (not derivable) in $\mathcal{N}^{\neg r}$ and $\mathcal{N}^{\neg l}$. The relation between implication and negation in the system $\mathcal{N}^{\neg r}$ is similar to the one in [3] and [4], the former unaware of the latter.

As is well known, characteristics of *Connexive Logics* (see [21] for a general survey; see also [16]) are the following (formal) theorems, which are *not* theorems of classical logic.

$$A_1 : \; \vdash \neg(\varphi \rightarrow \neg\varphi)$$
$$A_2 : \; \vdash \neg(\neg\varphi \rightarrow \varphi) \tag{1.1}$$

Both are jointly known as *Aristotle's thesis*. In [12], the other characteristic relationships

$$B_1 : \; (\varphi \rightarrow \psi) \rightarrow \neg(\varphi \rightarrow \neg\psi)$$
$$B_2 : \; (\varphi \rightarrow \neg\psi) \rightarrow \neg(\varphi \rightarrow \psi) \tag{1.2}$$

are attributed to the ancient philosopher and logician Boethius. For the history of connexive logics see [13].

However, the same intuition leading to Boethius' theses B_1 and B_2 leads also to the implications

$$B_3 : \; (\varphi \rightarrow \psi) \rightarrow \neg(\neg\varphi \rightarrow \psi)$$
$$B_4 : \; (\neg\varphi \rightarrow \psi) \rightarrow \neg(\varphi \rightarrow \psi) \tag{1.3}$$

as well as to the converses of B_i, $i = 1, \cdots, 4$. For B_3, see a derivation in 'Connexive Gentzen' ([13, p. 968], using Polish prefix notation).

Having *all* those B_is and their converses live together, and having implication transitive, leads to certain undesired complications, related to introducing and eliminating the operators (cf. the remark on p. 18). Identifying the two negations and creating one negation having the properties of both would:

- blur the distinction between the views of implication as focusing on sufficiency in contrast to focussing on necessity, a driving force behind the proposed systems. In particular,[1] $\varphi\to\psi$ becomes both necessary and sufficient for $\neg\varphi\to\neg\psi$, as shown by the following derivations.

$$\frac{\dfrac{\overline{(\varphi\to\psi)\to\neg(\neg\varphi\to\psi)}\ (B3)}{\neg(\neg\varphi\to\psi)\to(\neg\varphi\to\neg\psi)}\ (conv.\ B2)}{(\varphi\to\psi)\to(\neg\varphi\to\neg\psi)}\ (Trans\to) \qquad \frac{\dfrac{\overline{(\neg\varphi\to\neg\psi)\to\neg(\varphi\to\psi)}\ (B4)}{\neg(\varphi\to\neg\psi)\to(\varphi\to\psi)}\ (conv.\ B1)}{(\neg\varphi\to\neg\psi)\to(\varphi\to\neg\psi)}\ (Trans\to)$$

This equivalence does not conform with the standard meanings of sufficiency and necessity.

- render negation *ambiguous*, certainly an undesired effect.

- render negation *disharmonious* (see (4.41) below), also undesired if *Proof-Theoretic Semantics* (PTS) is adhered to, as I believe should be the case. See more on this in Section 4.

Therefore, I "split" the negation into two[2] negations, '$\neg_l$' and '$\neg_r$', each separately responsible to one of the A_i and two of the B_is, reformulated as formal theorems in terms of the two negations.

Below, all those characteristics are shown as formal theorems either of the ND-system $\mathcal{N}^{\neg_r}$ or of $\mathcal{N}^{\neg_l}$.

When viewed from a model-theoretic perspective, neither of those negations here is a *contradiction-forming* operator, except when applied to atomic propositions (justified below in Section 2.3). Rather, both are sub-contrariety formers (in two different ways). They play two separate roles for compound statements of the form $\alpha\to\beta$, distinguished as described below. Recall that a generic implication $\alpha\to\beta$ can be read in two ways.

- α is sufficient for β.

- β is necessary for α.

The two negations negate those two readings in the way described below.

---

[1] I thank Heinrich Wansing for this observation.
[2] Note that while I use the same technical term, split negation here is unrelated to the split negation in e.g., [19], the latter resulting from non-commutativity of the logic.

- From the NL point of view, '$\neg_r$' is a "corrective negation", expressing disagreement about *sufficiency* of the condition α, taking it sufficient for $\neg\beta$ instead of being sufficient for β.

- From the NL point of view, '$\neg_l$' too is a "corrective negation", expressing disagreement about *necessity* of the condition β for α, taking instead β as necessary for $\neg\alpha$.

Thus, the A_is express the impossibility of φ to be either necessary or sufficient for its own negation. If necessity and sufficiency are endowed a non-truth-functional meaning, one based on contents, then this interpretation of implication and negation expresses relationships between sentential meanings transcending the simple classical truth-functionality of implication and negation.

In the sequel, I consider a propositional fragment containing, in addition to atomic propositions, implication and negation only.

2 Natural language motivation

2.1 Negating implications

The point of departure is the following schematic dialog D between two participants A and B using two *compound* formulas (i.e., non-atomic, headed by a generic implication '$\rightarrow$') α, β in the following way:

$$D :: \begin{array}{l} A: \alpha \\ B: \text{No! } \beta \end{array} \tag{2.4}$$

Here α is $\varphi\rightarrow\psi$ (for some φ, ψ), while β is either $(\neg\varphi)\rightarrow\psi$ or $\varphi\rightarrow\neg(\psi)$. At this stage '$\neg$' is also considered a generic negation, to be made specific below.

The intended reading of the dialog D is characterised by the following two characteristics:

1. Participant B, by using No, *partially disagrees* with A about α by "negating" the latter.

2. Participant B offers β as the negated α expressing[3] the disagreement and "correcting" it.

Note that the "corrections" express consent about one argument of '$\rightarrow$', while negating the other argument of '$\rightarrow$'. I will refer to the arguments of '$\rightarrow$'s the left and right arguments, which explains the labels of the two negations. Clearly, this way of negating, *not* by contradicting, excludes the intuitionistic way of defining negation as implying $\perp$, absurdity.

I also consider a kind of a dual dialog, in which one of the arguments of '$\rightarrow$' is already negated. That is, α is $(\neg\varphi)\rightarrow\psi$ or $\varphi\rightarrow\neg(\psi)$ (for some φ, ψ), while β is, respectively, $\varphi\rightarrow\psi$. This suggest that double-negation elimination is employed in the "correction".

In the sequel, '$\vdash$' refers to derivability in the natural-deduction system $\mathcal{N}^{\neg r}$ and $\mathcal{N}^{\neg l}$, to be presented below, and $\dashv\vdash$ to mutual derivability. The context determines which proof-system is intended.

Metavariables φ, ψ range over compound formulas of some object language, and p, q over atomic propositions. I will first consider the negation of compound formulas, deferring to Section 2.3 the definition of negating an atomic proposition, using just $\neg p$ (unsubscripted) for its expression.

The ND-system $\mathcal{N}^{\neg r}$ and $\mathcal{N}^{\neg l}$ below induce the following mutual derivabilities, which can be interpreted as manifesting the sub-contrariness formation by the two negations.

$$\neg_l(\varphi\rightarrow\psi) \;\dashv\vdash\; \neg_l\varphi\rightarrow\psi \qquad \neg_r(\varphi\rightarrow\psi) \;\dashv\vdash\; \varphi\rightarrow\neg_r\psi \qquad (2.5)$$

The inspiration from the natural language dialog D pertains more to the first-degree case (without nesting of implications), but the incorporation of the generalisation with unrestricted nesting into the object language is needed in order to obtain a logic.

Before turning to a general theory, I will consider in some detail some instances of the dialog D, to get a better intuition about what is involved in partial disagreements

---

[3] In a naturally occurring dialog of the type D, a certain focal stress might be required. I ignore here such matters.

of the intended type. In Section 2.2 I correlate the D-dialogs to Ramsey's test and a newly considered *dual* of this test.

Example 2.1. *The first example I consider features an interaction between negation and a conditional, studied in detail by Cantwell in [3] under the name of 'conditional negation'. Cantwell's motivation is completely different. His intention is to remove the feature of the material implication of yielding a truth-value (actually, yielding the value 'true') in case the antecedent of the conditional yields a truth-value 'false'. The central feature of the interaction between negation and the conditional is the satisfaction of the following relation*

$$\neg(\varphi \rightarrow \psi) \vdash \varphi \rightarrow \neg\psi \tag{2.6}$$

This reflects a common view that the truth of an antecedent of a conditional is a *presupposition* of asserting that conditional. It remains a presupposition also of the assertion of the negated conditional.

I will strengthen this relation, in accord with (2.5), into

$$\neg(\varphi \rightarrow \psi) \dashv\vdash \varphi \rightarrow \neg\psi \tag{2.7}$$

below. Such a relationship, with a biconditional instead of mutual provability, is essential to the modal connexive logic introduced by Wansing (see [20, p. 371]).

The way the connectives are defined by Cantwell and made to satisfy (2.6) is via a model theory based on a certain three valued logic. Without being aware, he uses *the same* three-valued truth-tables for implication and negation as does [4].

His exemplary dialog featuring this interaction (not structured as D) is the following.
Anne: If Oswald didn't kill Kennedy, Jack Ruby did.
Bill: No! You're wrong.
Here is what Cantwell says ([3, p. 246]) about this exchange:

> When Bill denies the conditional asserted by Anne, he neither asserts nor denies that Oswald did the killing (he can continue, "If Oswald didn't kill Kennedy, Castro did"); his denial seemingly amounts to no more than the assertion that if Oswald didn't shoot Kennedy then neither did Jack Ruby. This kind of "conditional denial" seems to be a basic move in the language game; conditional negation is the sentential operator that

corresponds to this form of conditional denial: "It is not the case that if Oswald didn't shoot Kennedy, Jack Ruby did."

Not mentioned by Cantwell, this interaction between a conditional and a negation is one of the characteristics of connexive logics mentioned above. Yet another use of this way of negating an implication, not related to connexivity, is that by Dummett [6], confining it to negating a *subjunctive conditional*.

As I stated above, I want to approach the whole topic proof-theoretically, with no reference to truth-values, neither two nor any other number of them, or to relational frame semantics.

The examples below all use negated atomic propositions only, their negation understood informally by now (to be presented in more detail in Section 2.3).

Example 2.2. *Let us now consider a dialog structured as D, featuring a negated conditional. Suppose participants A and B are fans of the same soccer team T, but have opposing opinions as to how well T is prepared to play in a bad weather.*

$$D_1 :: \quad \begin{array}{l} A : \text{ If it rains, } T \text{ will win} \\ B : \text{ No! If it rains, } T \text{ will not win} \end{array} \tag{2.8}$$

That is, B consents about it raining, but disagrees as to what is raining a sufficient condition for. Considering (2.6) as reflecting this instance of D is best presented as

$$\neg_r(p {\rightarrow} q) \dashv\vdash p {\rightarrow} \neg_r q \tag{2.9}$$

In the dual dialog, we have

$$\hat{D}_1 :: \quad \begin{array}{l} B : \text{ If it rains, } T \text{ will not win} \\ A : \text{ No! If it rains, } T \text{ will win} \end{array} \tag{2.10}$$

represented as

$$\neg_r(p {\rightarrow} \neg_r q) \dashv\vdash p {\rightarrow} q \tag{2.11}$$

Here the effect of double negation elimination is manifested.

Example 2.3. *Similar arguments, motivated by the way implication and negation interact, are put forward in [4].*

Consider another instance of D between the same participants, the fans of team T.

$$D_2 :: \quad \begin{array}{l} A : \text{ If it rains, } T \text{ will win} \\ B : \text{ No! If it does not rain, } T \text{ will win} \end{array} \qquad (2.12)$$

Here B consents to team T winning, but disagrees about what the sufficient condition for that is. This can be modelled by[4]

$$\neg_l(p{\to}q) \dashv\vdash \neg_l p {\to} q \qquad (2.13)$$

Again, in the dual dialog, we have

$$\hat{D}_2 :: \quad \begin{array}{l} B : \text{ If it does not rain, } T \text{ will win} \\ A : \text{ No! If it rains, } T \text{ will win} \end{array} \qquad (2.14)$$

represented as

$$\neg_l(\neg_l p{\to}q) \dashv\vdash p {\to} q \qquad (2.15)$$

Here too is the effect of double negation elimination manifested.

In both the examples above, No expresses a way of negating a conditional different from the standard way of negating the material implication.

2.2   Ramsey's test and a dual test

In [17, p. 155], Ramsey proposes the following argument (quoted below with a slight modification of notation to fit the current presentation) as an interpretation of negating the conditional in NL.

> If two people are arguing 'If φ will ψ?' and are both in doubt as to φ, they are adding φ hypothetically to their stock of knowledge and arguing on that basis about ψ; so that in a sense 'If φ, ψ' and 'If φ, $\neg\psi$' are contradictories.

---

[4]This kind of conditional is not considered by Cantwell.

The scenario described in the above paragraph fits the structure of the argument between participants A and B in dialog D_1 (cf. (2.8)) used to motivate $\neg_r$. It exactly reflects an argument as to what is φ sufficient for: ψ or $\neg\psi$.

The connection to Ramsey's test was noted also by Ferguson [7].

I suggest a *dual test* with a scenario fitting the argument D_2 (cf. (2.12)), used to motivate $\neg_l$, that reflects an argument between A and B as to for which of φ, $\neg\varphi$ is ψ necessary for.

A dual Ramsey test:

> If two people are arguing 'If φ will ψ?' and are both in doubt as to φ, they are adding $\neg\varphi$ hypothetically to their stock of knowledge and arguing on that basis about ψ; so that in a sense 'If φ, ψ' and 'If $\neg\varphi$, ψ' are contradictories.

This dual test exactly reflects an argument as to what is ψ necessary for, φ or $\neg\varphi$.

2.3 Negating atomic propositions

Since atomic propositions are not implications, considerations like distinguishing between focus on sufficient conditions and necessary conditions do not apply to them and cannot drive the definition of their negation.

Consider a dialogue $\hat{\mathcal{D}}$, structured similarly to $\mathcal{D}$ (cf. 2.4)),where α and β are both atomic propositions.

Example 2.4 (atomic dialog:).

$$\hat{\mathcal{D}} :: \quad \begin{array}{l} A : \ T \text{ will win} \\ B : \ \text{No! } T \text{ will not win} \end{array}$$

Here participant B plainly disagrees with participant A's assessment about the outcome of a game involving team T, not involving any conditionality. Here B's correction of A's statement (following his NO!) is just a claim of the opposite proposition, clearly attempting to contradict A.

This example exemplifies the idea behind defining

$$\neg_l p = \neg_r p = \neg p \tag{2.16}$$

(where $\neg$ is classical negation). This will lead to the ND-rules for atomic proposition in Section 3 to coincide with the classical ones.

3 The natural-deduction system $\mathcal{N}^{\neg_r}$ and $\mathcal{N}^{\neg_l}$

The design of the ND-systems $\mathcal{N}^{\neg_r}$ and $\mathcal{N}^{\neg_l}$ in Figures 1 and 2, respectively, is based on the following principles.

1. Both negations, as mentioned above, are not contradiction-forming, except when applied to atomic propositions. Rather, both $\neg_r(\varphi \rightarrow \psi)$ and $\neg_l(\varphi \rightarrow \psi)$ are sub-contraries of $\varphi \rightarrow \psi$. This behaviour is very similar to what is known in the semantics of natural language as 'neg raising' (see [10] for discussion and references).

2. The negations and the implication are not *independent*, and have to be understood *together*. Technically, this means that the I/E-rules for the implication are *not pure* (i.e., refer to more than one operator) [6]. In the model-theory in Section 5, the dependence between negations and implication results in a non-compositionality in the assignment of truth-value.

3. Negations here are *non-uniform*, their (I/E)-rules depending on the negated formula. There are no rules that might be seen as '$(\neg_r I)$' and '$(\neg_l I)$' by which $\neg_r \varphi$ and $\neg_l \varphi$ can be introduced for a "bare" compound φ. The negations $\neg_r, \neg_l$ can only be introduced for an implication, and this can be done in two ways. Accordingly, the two negated implications are eliminated differently. This is a major need for splitting the negation, as will become even clearer in Section 4.

   One result of this non-uniformity of negation is that both the ND-systems introduced below do not admit the rule of *uniform substitution*. Atomic propositions are *not* propositional variables. This is also reflected in the definition of assignments in the model theory (see Definition 5.4).

4. The two systems $\mathcal{N}^{\neg_r}$ and $\mathcal{N}^{\neg_l}$ cannot conveniently be amalgamated into one combined system. There is an issue of how to propagate negation to the

appropriate argument of the implication. Suppose one considers (in an alleged combined system) $\neg_r(\varphi\to\psi)$ by negating ψ. Which nested negation should be employed? In principle, both ways can do, leading both to $\varphi\to\neg_l\psi$ and $\varphi\to\neg_r\psi$. A similar situation pertains to $\neg_l(\varphi\to\psi)$, leading either to $\neg_r\varphi\to\psi$ or to $\neg_l\varphi\to\psi$. This does not lead to a coherent interpretation of the A_is and B_is, that should relate to *one and the same negation* each. This reinterpretation of the A_is and B_is is seen clearly in the separate systems.

5. While iterating the same negation makes, giving rise to two forms of double negation, both eliminable, the iterations

$$\neg_r\neg_l\varphi, \qquad \neg_l\neg_r\varphi \tag{3.17}$$

do not seem to have an obvious interpretation. Those iterations are not well-formed if separation of systems is kept.

Additional remarks about the I/E-rules:

1. Classically, the double-negation I/E-rules are related to the reversing of truth value associated with a contradiction-forming operator. Here, they originate from a different source. As negations are associated with disagreement about one of the arguments of '$\to$', when applied to an already negated argument, a negation finds, so to speak, nothing (that is, no implication) to disagree about, so it cancels the disagreement when eliminated. When introduced, it can be seen, so to speak, as retracting the disagreement that would have been formed by negating once only.

2. Note that the '(dni)' rules are *primitive*, in contrast to classical '(dni)', which is derivable in classical logic; this is again an effect of the current negations not being contradiction-forming, blocking the usual classical derivation of '(dni)'.

A methodological remark: In what sense can $\neg_r$ and $\neg_l$ "deserve" to be considered as negations? First, both are *involutive*, as is common for several other negations. Secondly, they are formed by a tool not used before in proof-theory: *negating a rule* (in contrast to the usual notion of negating a proposition). Here the rule classical/intuitionistic $(\to I)$, introducing an implication, is negated in two ways:

1. Negating the discharged assumption of the premise.

$$
\frac{\begin{array}{c}[\varphi]_i\\ \vdots\\ \psi\end{array}}{\varphi\to\psi}\ (\to I^i) \qquad \frac{\varphi\to\psi \quad \varphi}{\psi}\ (\to E) \tag{3.18}
$$

$$
\frac{\begin{array}{c}[\varphi]_i\\ \vdots\\ \neg_r\psi\end{array}}{\neg_r(\varphi\to\psi)}\ (\neg_r\to I^i) \qquad \frac{\neg_r(\varphi\to\psi) \quad \varphi}{\neg_r\psi}\ (\neg_r\to E) \tag{3.19}
$$

$$
\frac{\begin{array}{cc}[p]_i & [p]_i\\ \vdots & \vdots\\ q & \neg_r q\end{array}}{\neg_r p}\ (At\neg_r I^i) \qquad \frac{p \quad \neg_r p}{\varphi}\ (At\neg_r E) \tag{3.20}
$$

$$
\frac{\neg_r\neg_r\varphi}{\varphi}\ (dne_r) \qquad \frac{\varphi}{\neg_r\neg_r\varphi}\ (dni_r) \tag{3.21}
$$

Figure 1: The I/E-rules of $\mathcal{N}^{\neg r}$

2. Negating the conclusion of the sub-derivation forming the premise.

As for $(\to E)$, it is negated either by negating its minor premise or by negating its conclusion. Similar rules, but in a sequent calculus L/R-rules form, appear in [11], but are not viewed as negating the standard L/R rules for implication.

Note that both $\mathcal{N}^{\neg r}$ and $\mathcal{N}^{\neg l}$ are *paraconsistent*, invalidating explosion (cf. Example 5.7). Neither one of $\varphi, \neg_r\varphi \vdash \psi$ and $\varphi, \neg_l\varphi \vdash \psi$ holds (except for an atomic φ). This is typical to sub-contrariety forming operators [2].

3.1   The ND-system $\mathcal{N}^{\neg r}$

Derivations (tree-shaped) are defined recursively as usual.

Proposition 3.1 (closure under composition). *Derivations in $\mathcal{N}^{\neg r}$ are closed under composition of derivations.*

The proof is standard and omitted.

Corollary 1. *The mutual derivability* $\neg_r(\varphi{\rightarrow}\psi)$ $\dashv\vdash$ $\varphi{\rightarrow}\neg_r\psi$ *(cf. (2.5)) holds.*

Proof: The derivations are as follows.

$$\cfrac{\cfrac{\varphi{\rightarrow}\neg_r\psi \quad [\varphi]_1}{\neg_r\psi}\ (\rightarrow E)}{\neg_r(\varphi{\rightarrow}\psi)}\ (\neg_r I^1) \qquad \cfrac{\cfrac{\neg_r(\varphi{\rightarrow}\psi) \quad [\varphi]_1}{\neg_r\psi}\ (\neg_r\rightarrow E)}{\varphi{\rightarrow}\neg_r\psi}\ (\neg_r\rightarrow I^1) \tag{3.22}$$

3.1.1 Some properties of $\mathcal{N}^{\neg r}$

In this section, I show some of the properties of the $\mathcal{N}^{\neg r}$ system, justifying its being connexive.

Proposition 3.2 (Aristotle's $\neg_r$-thesis).

$$\vdash \neg_r(\varphi{\rightarrow}\neg_r\varphi) \tag{3.23}$$

Proof: The derivation of is as follows.

$$\cfrac{\cfrac{[\varphi]_1}{\neg_r\neg_r\varphi}\ [(dni_r)]}{\neg_r(\varphi{\rightarrow}\neg_r\varphi)}\ (\neg_r\rightarrow I^1)$$

Proposition 3.3 (Boethius' $\neg_r$-theses).

$$(B1) \vdash (\varphi{\rightarrow}\psi){\rightarrow}\neg_r(\varphi{\rightarrow}\neg_r\psi)$$

$$(B2) \vdash (\varphi{\rightarrow}\neg_r\psi){\rightarrow}\neg_r(\varphi{\rightarrow}\psi) \tag{3.24}$$

Proof: The derivation of (B1) is as follows.

$$\cfrac{\cfrac{\cfrac{\cfrac{[\varphi{\rightarrow}\psi]_2 \quad [\varphi]_1}{\psi}\ (\rightarrow E)}{\neg_r\neg_r\psi}\ (dni_r)}{\neg_r(\varphi{\rightarrow}\neg_r\psi)}\ (\neg_r\rightarrow I^1)}{(\varphi{\rightarrow}\psi){\rightarrow}\neg_r(\varphi{\rightarrow}\neg_r\psi)}\ (\rightarrow I^2) \tag{3.25}$$

491

The derivation of (B2) is as follows.

$$\frac{\dfrac{\dfrac{[\varphi{\to}\neg_r\psi]_2 \quad [\varphi]_1}{\neg_r\psi}\,(\to E)}{\dfrac{\neg_r(\varphi{\to}\psi)}{}\,(\neg_r{\to}I_2^1)}}{(\varphi{\to}\neg_r\psi){\to}\neg_r(\varphi{\to}\psi)}\,(\to I^2) \tag{3.26}$$

Example 3.5.

$$\neg_r(\varphi{\to}(\psi{\to}\chi))\vdash(\varphi{\to}(\psi{\to}\neg_r\chi))$$

The derivation is

$$\frac{\dfrac{\dfrac{\neg_r(\varphi{\to}(\psi{\to}\chi))}{(\varphi{\to}\neg_r(\psi{\to}\chi))}\,(2.5)}{(\varphi{\to}(\psi{\to}\neg_r\chi))}\,(2.5)}{}$$

3.2 The natural-deduction system $\mathcal{N}^{\neg_l}$

The design of the ND-system $\mathcal{N}^{\neg_l}$ in Figure 2 is based on analogous principles to those driving $\mathcal{N}^{\neg_r}$.. *Derivations* (tree-shaped) are once again defined recursively as usual.

Proposition 3.4 (closure under composition). *Derivations of $\mathcal{N}^{\neg_l}$ are closed under composition of derivations.*

Again, the proof is standard and omitted.

Corollary 2. *The mutual derivability $\neg_l(\varphi{\to}\psi)$ $\dashv\vdash$ $\neg_l\varphi{\to}\psi$ (cf. (2.5)) holds.*

Proof: The derivations are as follows.

$$\frac{\dfrac{\dfrac{\neg_l\varphi{\to}\psi \quad [\neg_l\varphi]_1}{\psi}\,(\to E)}{\neg_l(\varphi{\to}\psi)}\,(\neg_l{\to}I^1)}{} \qquad \frac{\dfrac{\dfrac{\neg_l(\varphi{\to}\psi) \quad [\neg_l\varphi]_1}{\psi}\,(\neg_l{\to}E)}{\neg_l\varphi{\to}\psi}\,(\to I^1)}{} \tag{3.31}$$

3.2.1 Some properties of $\mathcal{N}^{\neg_l}$

In this section, I show some of the properties of the system, justifying its being connexive.

$$\frac{\begin{array}{c}[\varphi]_i \\ \vdots \\ \psi\end{array}}{\varphi{\rightarrow}\psi}\ (\rightarrow I^i) \qquad \frac{\varphi{\rightarrow}\psi \quad \varphi}{\psi}\ (\rightarrow E) \tag{3.27}$$

$$\frac{\begin{array}{c}[\neg_l\varphi]_i \\ \vdots \\ \psi\end{array}}{\neg_l(\varphi{\rightarrow}\psi)}\ (\neg_l{\rightarrow}I^i) \qquad \frac{\neg_l(\varphi{\rightarrow}\psi) \quad \neg_l\varphi}{\psi}\ (\neg_l{\rightarrow}E) \tag{3.28}$$

$$\frac{\begin{array}{cc}[p]_i & [p]_i \\ \vdots & \vdots \\ q & \neg_l q\end{array}}{\neg_l p}\ (At\neg_l I^i) \qquad \frac{p \quad \neg_l p}{\varphi}\ (At\neg_l E) \tag{3.29}$$

$$\frac{\neg_l\neg_l\varphi}{\varphi}\ (dne_l) \qquad \frac{\varphi}{\neg_l\neg_l\varphi}\ (dni_l) \tag{3.30}$$

Figure 2: The I/E-rules of $\mathcal{N}^{\neg_l}$

Proposition 3.5 (Aristotle's $\neg_l$-thesis).

$$\vdash \neg_l(\neg_l\varphi{\rightarrow}\varphi) \tag{3.32}$$

Proof: The derivation is as follows.

$$\frac{\dfrac{[\neg_l\neg_l\varphi]_1}{\varphi}\ [(dne_l)}{\neg_l(\neg_l\varphi{\rightarrow}\varphi)}\ (\neg_l{\rightarrow}I^1)$$

Proposition 3.6 (Boethius' $\neg_l$-theses).

$$(B3) \vdash (\varphi{\rightarrow}\psi){\rightarrow}\neg_l(\neg_l\varphi{\rightarrow}\psi) \tag{3.33}$$

$$(B4) \vdash (\neg_l\varphi{\rightarrow}\psi){\rightarrow}\neg_l(\varphi{\rightarrow}\psi)$$

Proof: The derivation of (B3) is as follows.

$$\cfrac{\cfrac{[\varphi\to\psi]_2 \quad \cfrac{[\neg_l\neg_l\varphi]_1}{\varphi}\ (dne_l)}{\psi}\ (\to E)}{\cfrac{\neg_l(\neg_l\varphi\to\psi)}{(\varphi\to\psi)\to\neg_l(\neg_l\varphi\to\psi)}\ (\neg_l\to I_1^1)}\ (\to I^2) \tag{3.34}$$

The derivation of (B4) is as follows.

$$\cfrac{\cfrac{\cfrac{[\neg_l\varphi\to\psi]_2 \quad [\neg_l\varphi]_1}{\psi}\ (\to E)}{\neg_l(\varphi\to\psi)}\ (\neg_l\to I^1)}{(\neg_l\varphi\to\psi)\to\neg_l(\varphi\to\psi)}\ (\to I^2) \tag{3.35}$$

4 Qualification of $\mathcal{N}^{\neg_r}$ and $\mathcal{N}^{\neg_l}$ as meaning conferring

According to the *proof-theoretic semantics* (PTS) programme (see [18] for a survey and [8] for a detailed presentation), meaning is determined by *canonical derivability conditions* in a meaning-conferring proof systems. Those conditions are based on *grounds for assertion*. This theory of meaning constitutes an alternative to *model-theoretic semantics* (MTS), identifying meaning as truth-conditions (in models of a suitable form).

As is well known, not every ND-system qualifies as meaning-conferring. One prominent criterion for such qualification requires a certain balance between the *I*-rules and the *E*-rules, none of those groups of rules out-powers the other. Technically, this balance is known as the conditions of *harmony* and *stability* [6].

4.1 Harmony

One of the formalisation of harmony is by means of the property of *local-soundness* [14],[5].

Definition 4.1 (maximal formula). *An occurrence of a formula φ in a derivation $\mathcal{D}$ is* maximal *iff it is the result of an application of an I-rule and a major premise of an E-rule.*

494

Definition 4.2 (local-soundness). *An ND-system $\mathcal{N}$ is* locally-sound *iff every $\mathcal{N}$-derivation $\mathcal{D}$ that has an occurrence of a maximal formula can be transformed into an equivalent derivation $\mathcal{D}'$ (i.e., having the same, or less, assumptions and the same conclusion) in which the maximal φ doe not occur. Such a transformation is called a* reduction.

Failure of local-soundness, i.e., the presence of a non-reducible maximal formula φ, indicates that the *I*-rules are too strong compared to the *E*-rules, yielding a conclusion not derivable without introducing φ.

To present reductions, the following notations are used. For a derivation $\mathcal{D}_1$ having an assumption φ, and a derivation $\mathcal{D}_2$ having a conclusion φ, the notation $\mathcal{D}$ :: $\mathcal{D}_1[\varphi := \overset{\mathcal{D}_2}{\varphi}]$ indicates the tree obtaining by replacing every leaf labeled φ in $\mathcal{D}_1$ by the tree $\mathcal{D}^2$ (identifying that leaf of $\mathcal{D}_1$ with the root of $\mathcal{D}_2$). $\mathcal{D}$ is a derivation whenever $\mathcal{N}$ is closed under derivation composition.

Proposition 4.7 (local-soundness of $\mathcal{N}^{\neg r}$ and $\mathcal{N}^{\neg l}$). *$\mathcal{N}^{\neg r}$ and $\mathcal{N}^{\neg l}$ are locally-sound.*

Proof: Below are the reductions establishing local-soundness. They are listed according to the *I*-rule generating a maximal formula. All the reductions are well-defined by Proposition 3.1.

$(\to I)$: This is the standard reduction for implication [15].

$$
\frac{\begin{array}{c}[\varphi]_i\\ \mathcal{D}_1\\ \psi\end{array}}{\varphi\to\psi}(\to I^i)\quad \begin{array}{c}\mathcal{D}_2\\ \varphi\end{array} \atop \frac{}{\psi}(\to E)\qquad \leadsto_r \qquad \mathcal{D}_1[\varphi := \overset{\mathcal{D}_2}{\varphi}] \atop \psi \qquad\qquad (4.36)
$$

$(\neg_l\to I)$: This is similar to the standard reduction for implication.

$$
\frac{\begin{array}{c}[\neg_l\varphi]_i\\ \mathcal{D}_1\\ \psi\end{array}}{\neg_l(\varphi\to\psi)}(\neg_l\to I^i)\quad \begin{array}{c}\mathcal{D}_2\\ \neg_l\varphi\end{array} \atop \frac{}{\psi}(\neg_l\to E)\qquad \leadsto_r \qquad \mathcal{D}_1[\neg_l\varphi := \overset{\mathcal{D}_2}{\neg_l\varphi}] \atop \psi \qquad\qquad (4.37)
$$

495

$(\neg_r {\rightarrow} I)$: This is also similar the standard reduction for implication.

$$\cfrac{\cfrac{\begin{array}{c}[\varphi]_i\\ \mathcal{D}_1\\ \neg_r\psi\end{array}}{\neg_r(\varphi{\rightarrow}\psi)}\;(\neg_r{\rightarrow}I^i)\quad \begin{array}{c}\mathcal{D}_2\\ \varphi\end{array}}{\neg_r\psi}\;(\neg_r{\rightarrow}E) \qquad \leadsto_r \qquad \cfrac{\mathcal{D}_1[\varphi := \begin{array}{c}\mathcal{D}_2\\ \varphi\end{array}]}{\neg_r\psi} \tag{4.38}$$

$At\neg_l I$:

$$\cfrac{\cfrac{\begin{array}{cc}[p]_1 & [p]_1\\ \mathcal{D}_1 & \mathcal{D}_2\\ q & \neg_l q\end{array}}{\neg_l p}\;(At\neg_l I^1)\quad \begin{array}{c}\mathcal{D}_3\\ p\end{array}}{\varphi}\;(At\neg_l E) \qquad \leadsto \qquad \cfrac{\begin{array}{cc}[p]_1 & [p]_1\\ \mathcal{D}_1 & \mathcal{D}_2\\ q & \neg_l q\end{array}}{\varphi}\;(At\neg_l E) \tag{4.39}$$

(dni_l):

$$\cfrac{\cfrac{\cfrac{\mathcal{D}_1}{\varphi}}{\neg_l\neg_l\varphi}\;(dni_l)}{\varphi}\;(dne_l) \qquad \leadsto \qquad \begin{array}{c}\mathcal{D}_1\\ \varphi\end{array} \tag{4.40}$$

(dni_r): Similar.

Remark: Here one can realise why negation had to be split. If we had just one '$\neg$', but with

- two $\neg{\rightarrow}I$-rules, one like $\neg_l{\rightarrow}I$ and one like $\neg_r{\rightarrow}I$

- and two $\neg{\rightarrow}E$-rules, one like $\neg_l{\rightarrow}E$ and one like $\neg_r{\rightarrow}E$

then the following derivation would be irreducible. It introduces like $\neg_l$ but eliminates like $\neg_r$.

$$\cfrac{\cfrac{\begin{array}{c}[\neg\varphi]_1\\ \mathcal{D}_1\\ \psi\end{array}}{\neg(\varphi{\rightarrow}\psi)}\;(\neg{\rightarrow}I_2)\quad \begin{array}{c}\mathcal{D}_2\\ \varphi\end{array}}{\neg\psi}\;(\neg{\rightarrow}E_2) \tag{4.41}$$

Such a system would be bluntly disharmonious.

4.2 Stability

This property constitutes the other "half" of the required balance between the I/E-rules for qualifying as meaning conferring. It is formalised (to a certain approximation) by *local-completeness* [14],[5].

Definition 4.3 (local-completeness). *An ND-system $\mathcal{N}$ is locally-complete iff every $\mathcal{N}$-derivation $\overset{\mathcal{D}}{\varphi}$ can be transformed to an equivalent derivation decomposing φ by E-rules and recomposing by I-rules. Such a transformation is called an* expansion.

Failure of local-completeness, i.e., the presence of a non-expandable derivation of some φ, indicates that the I-rules are too weak compared to the E-rules, yielding a conclusion not derivable without eliminating φ.

Proposition 4.8 (local-completeness of $\mathcal{N}^{\neg r}$ and $\mathcal{N}^{\neg l}$). *$\mathcal{N}^{\neg r}$ and $\mathcal{N}^{\neg l}$ are locally-complete.*

Proof: Below are the required expansions.

$\varphi{\rightarrow}\psi$: This is again a standard expansion for implication [15].

$$
\begin{array}{cc}
\begin{array}{c} \mathcal{D} \\ \varphi{\rightarrow}\psi \end{array} \leadsto_e &
\dfrac{\dfrac{\dfrac{\mathcal{D}}{\varphi{\rightarrow}\psi}\quad [\varphi]_1}{\psi}\,(\rightarrow E)}{\varphi{\rightarrow}\psi}\,(\rightarrow I^1)
\end{array}
\tag{4.42}
$$

$\neg_l(\varphi{\rightarrow}\psi)$:

$$
\begin{array}{cc}
\begin{array}{c} \mathcal{D} \\ \neg_l(\varphi{\rightarrow}\psi) \end{array} \leadsto_e &
\dfrac{\dfrac{\dfrac{\mathcal{D}}{\neg_l(\varphi{\rightarrow}\psi)}\quad [\neg_l\varphi]_1}{\psi}\,(\neg_l{\rightarrow}E)}{\neg_l(\varphi{\rightarrow}\psi)}\,(\neg_l{\rightarrow}I^1)
\end{array}
\tag{4.43}
$$

$\neg_r(\varphi{\rightarrow}\psi)$:

$$
\begin{array}{cc}
\begin{array}{c} \mathcal{D} \\ \neg_r(\varphi{\rightarrow}\psi) \end{array} \leadsto_e &
\dfrac{\dfrac{\dfrac{\mathcal{D}}{\neg_r(\varphi{\rightarrow}\psi)}\quad [\varphi]_1}{\neg_r\psi}\,(\neg_r{\rightarrow}E)}{\neg_r(\varphi{\rightarrow}\psi)}\,(\neg_r{\rightarrow}I^1)
\end{array}
\tag{4.44}
$$

$\neg_r \neg_r \varphi$:

$$
\frac{\mathcal{D}}{\neg_r \neg_r \varphi} \quad \rightsquigarrow_e \qquad
\frac{\dfrac{\dfrac{\mathcal{D}}{\neg_r \neg_r \varphi}}{\varphi}\,(dni_r)E}{\neg_r \neg_r \varphi}\,(dne_r)I
\tag{4.45}
$$

$\neg_l \neg_l \varphi$: Similar.

For an extensive discussion about the meaning determined by a qualified meaning-conferring ND-system see [9].

5   Model-theory for $\mathcal{N}^{\neg r}$ and $\mathcal{N}^{\neg l}$

In this section, I present a model-theory for the two ND proof-systems $\mathcal{N}^{\neg r}$ and $\mathcal{N}^{\neg l}$ presented above. Let me stress again that *I do not intend this model-theory to serve as a semantics* for the induced logic. As I stated before, I see the ND-systems as *meaning-conferring*, a definitional tool. The role of the model-theory is merely a tool for establishing indirectly some properties of the ND-systems, such as *non-derivability*.

The model theory of both systems is based on a *four-valued* system, having the values $\{0, 1, 2, 3\}$. As an intuitive handle to the interpretation of those values, one can think of them as binary representations of *ordered pairs* of classical truth values $\{0, 1\}$. Let the *designated* values be $D = \{0, 1, 3\}$. Let me mention that this model-theory is not related to the ND-systems $\mathcal{N}^{\neg r}$ and $\mathcal{N}^{\neg l}$ in accordance to a general pattern relating multi-valued logics to their corresponding ND-system as specified in [1].

The characteristic fact of the definition of the two negations is their *non-compositionality*: they are not truth-functional in that the value of a negated implication does not depend on the value of the implication itself; rather, it depends on the values of both the antecedent and the consequent of the implication, coded as an ordered pair of classical values. Thus, the value 2, coding $\langle 1, 0 \rangle$, is the falsity in this system, coding the value of an implication with a true antecedent and a false consequent. In that, the model-theory reflects the *non-purity* of the negated-implication rules in both systems. The other values are variants of classical truth, recording *in*

p	$\neg_r p$	$\neg_l p$
2	3	3
3	2	3

φ	$\neg_r \varphi$	$\neg_l \varphi$
3	1	2
2	0	3
1	3	0
0	2	1

Figure 3: The truth-tables for $\neg_r$ and $\neg_l$

virtue of which combination of values for the consequent and the antecedent is the implication true.

When considering truth-assignments, atomic propositions differ from compound propositions (implications): the former are assigned *only* $\{2, 3\}$, rendering them contradictory.

Definition 5.4 (assignment). *An assignment σ is a mapping satisfying*

$$\begin{cases} \sigma[\![p]\!] \in \{2, 3\} & atomic \\ \sigma[\![\varphi]\!] \in \{0, 1, 2, 3\} & compound \end{cases}$$

Thus, neither of the logics is closed under the rule of uniform substitution. This reflects an intuition coming from natural language, that the generators of the object language are *atomic propositions*, having unspecified fixed contents, and not propositional variables! A similar phenomenon, similarly justified, takes place also in the system of [4].

The truth tables for the two negations are presented in Figure 3. The truth-table of the implication is presented in Figure 4. The line marked with $(*)$ are the falsity

φ	ψ	$\varphi\rightarrow\psi$	
3	3	3	
3	2	2	$(*)$
3	1	3	
3	0	3	
2	3	1	
2	2	0	
2	1	1	
2	0	1	
1	3	3	
1	2	2	$(*)$
1	1	3	
1	0	3	
0	3	3	
0	2	2	$(*)$
0	1	3	
0	0	3	

Figure 4: The truth-table for $\rightarrow$

lines, yielding false for a false consequent and the various ways the antecedent can be true (designated).

Definition 5.5 (equivalence). *Let $\varphi \equiv \psi$ iff for every assignment σ: $\sigma[\![\varphi]\!] = \sigma[\![\psi]\!]$.*

In those systems, *soundness* of a rule means that for every assignment σ: if the premises of the rule are assigned a designated value under σ, so does the conclusion. A simple case analysis establishes the following proposition.

Proposition 5.9 (soundness). $\mathcal{N}^{\neg r}$ *and* $\mathcal{N}^{\neg l}$ *are sound w.r.t. the model-theory.*

The proof is by case analysis. As an example, consider the rule $(\neg_l \rightarrow I)$ (in Figure 2). In order to lead from a designated value for $\neg_l \varphi$ to a designated value of ψ, the possible values of φ are $\{1, 0, 2\}$ and the possible values of ψ are $\{0, 1, 3\}$. For each combination of the above, the value of $\neg_l(\varphi \rightarrow \psi)$ is 0, designated too.

Similarly, the following equivalences (identical truth-tables) justify (2.5).

$$\neg_r(\varphi \rightarrow \psi) \equiv \varphi \rightarrow \neg_r \psi \qquad \neg_l(\varphi \rightarrow \psi) \equiv \neg_l \varphi \rightarrow \psi$$

Example 5.6 (non-derivability of contraposition). *As an example of non-derivability, both negations invalidate* contraposition *(as might be expected).*

$\neg_r$: *Consider the assignment σ' under which $\sigma'[\![\psi]\!] = 2$ and $\sigma'[\![\varphi]\!] = 3$. Hence, $\sigma'[\![\neg_r \psi]\!] = 3$ and $\sigma'[\![\neg_r \varphi]\!] = 2$. Therefore, $\sigma'[\![\neg_r \psi \rightarrow \neg_r \varphi]\!] = 2$ (false!), while $\sigma'[\![\varphi \rightarrow \psi]\!] = 3$ (true!).*

$\neg_l$: *Consider the assignment σ'' under which $\sigma''[\![\psi]\!]i = 1$ and $\sigma''[\![\varphi]\!] = 0$. Hence, $\sigma''[\![\neg_l \psi]\!] = 3$ and $\sigma''[\![\neg_l \varphi = 2]\!]$. Therefore, $\sigma''[\![\neg_l \psi \rightarrow \neg_l \varphi]\!] = 2$ (false!), while $\sigma''[\![\varphi \rightarrow \psi]\!] = 3$ (true!).*

Contraposition does hold for atomic sentences, for which such assignments are inadmissible. The same situation obtains also in the system of [4].

As another example, I show non-explosion.

Example 5.7 (non-explosiveness). *Consider an assignment σ^* s.t. $\sigma^*[\![\varphi]\!] = \sigma^*[\![\psi]\!] = 3$. Thus, both $\sigma^*[\![\neg_l(\varphi \rightarrow \psi)]\!] [= \sigma^*[\![\neg_l \varphi \rightarrow \psi]\!]] = 3$ (truth!), and $\sigma^*[\![\varphi \rightarrow \psi]\!] = 3$ (truth), but an arbitrary ξ can certainly have $\sigma^*[\![\xi]\!] = 2$ (falsity!). Thus,*

$$\neg_l(\varphi \rightarrow \psi), (\varphi \rightarrow \psi) \not\models \xi$$

A similar argument applies to '$\neg_r$'.

Example 5.8 (Modus Tollens). *Neither of the two versions of the modus tollens rule are validated.*

$$\frac{\varphi \to \psi \quad \neg_r \psi}{\neg_r \varphi} \ (MT_r) \qquad \frac{\varphi \to \psi \quad \neg_l \psi}{\neg_l \varphi} \ (MT_l) \tag{5.46}$$

- *A counter example for (MT_r): consider an assignment σ with $\sigma[\![\varphi]\!] = 0$ and $\sigma[\![\psi]\!] = 3$. Therefore, $\sigma[\![\varphi \to \psi]\!] = 3$ (designated), and $\sigma[\![\neg_r \psi]\!] = 1$ (designated); however, $\sigma[\![\neg_r \varphi]\!] = 2$ (non-designated).*

- *A counter example for (MT_l): consider an assignment σ with $\sigma[\![\varphi]\!] = 3$ and $\sigma[\![\psi]\!] = 1$. Therefore, $\sigma[\![\varphi \to \psi]\!] = 3$ (designated), and $\sigma[\![\neg_l \psi]\!] = 0$ (designated); however, $\sigma[\![\neg_l \varphi]\!] = 2$ (non-designated).*

6 Conclusions

In this paper, I have introduced two negations, interacting with implication in a manner similar to the interaction in some known Connexive Logics. This interaction is inspired by a similar interaction present in some natural language dialogs, where negation is used to disagree about sufficiency or necessity of conditions, "correcting" the negated proposition by an alternative proposition, in which either the antecedent or the consequent of the negated implication are negated themselves.

Two ND proof-systems are proposed for the resulting logics, viewed as meaning-conferring, in accordance with the proof-theoretic semantics programme. In addition, a four-valued model-theory is developed as a tool for establishing non-derivability.

An interesting alternative for capturing the same intuition might be keeping *one* negation, but splitting the implication. I leave this for further research. Also left for future development is the extension by adding other connectives, where again negation applies by negating one argument of a binary connective. For conjunction and disjunction, this would produce an invalidation of De Morgan's rules. Negating quantified propositions by such negations is, of course, of interest too.

References

[1] M. Baaz, C. G. Fermüller, and R. Zach. Systematic construction of natural deduction systems for many-valued logics. In *Proceedings of the 23rd International Symposium on Multiple Valued Logic*, pages 208–213, Los Alamos, CA, 1993. IEEE Computer Society Press.

[2] J.-Y. Béziau. Paraconsistent logic and contradictory viewpoints. *Revista Brasileira de Filosofia*, (242), 2015. to appear.

[3] J. Cantwell. The logic of conditional negation. *Notre Dame Journal of Formal Logic*, 49(3):245–260, 2008.

[4] W. S. Cooper. The propositional logic of ordinary discourse 1. *Inquiry*, 11(1–4):295–320, 1968.

[5] R. Davies and F. Pfenning. A modal analysis of staged computation. *Journal of the ACM*, 48(3):555—604, 2001.

[6] M. Dummett. *The Logical Basis of Metaphysics*. Harvard University Press, Cambridge, MA, 1991.

[7] T.M. Ferguson. Ramsey's footnote and Priest's connexive logics. *Bulletin of Symbolic Logic*, 20(3):387–388, 2014.

[8] N. Francez. *Proof-Theoretic Semantics*. College Publications, London, 2015.

[9] N. Francez. Views of proof-theoretic semantics: Reified proof-theoretic meanings. *Journal of Logic and Computation*, 26(2):479–494, 2016.

[10] L. R. Horn and H. Wansing. Negation. In E. N. Zalta, editor, *The Stanford Encyclopedia of Philosophy*. Summer 2015 edition, 2015.

[11] N. Kamide and H. Wansing. Connexive modal logic based on positive *S4*. In J.-Y. Béziau and M. Coniglio, editors, *Logic without Frontiers*, pages 389–409. College Publications, London, 2011.

[12] W. Kneale and M. Kneale. *The Development of Logic*. Oxford University Press, Oxford, 1962.

[13] S. McCall. A history of connexivity. In D. Gabbay et al., editors, *A History of Logic*, volume 11, pages 415–449. Elsevier, Amsterdam, 2012.

[14] F. Pfenning and R. Davies. A judgmental reconstruction of modal logic. *Mathematical Structures in Computer Science*, 11(4):511–540, 2001.

[15] D. Prawitz. *Natural Deduction: A Proof-Theoretical Study*. Almqvist and Wicksell, Stockholm, 1965.

[16] G. Priest. Negation as cancellation and connexive logic. *Topoi*, 18(2):141–148, 1999.

[17] F. P. Ramsey. General propositions and causality. In D. H. Mellor, editor, *Philosophical Papers*, pages 145–163. Cambridge University Press, Cambridge, 1990.

[18] P. Schroeder-Heister. Proof-theoretic semantics. In E. N. Zalta, editor, *The Stanford Encyclopedia of Philosophy*. Spring 2016 edition, 2016.

[19] S. Sequoiah-Grayson. Dynamic negation and negative information. *Review of Symbolic*

Logic, 2(1):233–248, 2009.

[20] H. Wansing. Connexive modal logic. In R. Schimdt, I. Pratt-Hartmann, M. Reynolds, and H. Wansing, editors, *Advances in Modal Logic*, volume 5, pages 367–383. College Publications, London, 2005.

[21] H. Wansing. Connexive logic. In E. N. Zalta, editor, *The Stanford Encyclopedia of Philosophy*. Fall 2015 edition, 2015.

Received October 2015

A Note on Francez' Half-Connexive Formula

Hitoshi Omori*
Department of Philosophy
Kyoto University, Japan
hitoshiomori@gmail.com

Abstract

The present note examines an unusual formula studied by Nissim Francez. More specifically, a variant of Nelson's logic is introduced along the lines of the connexive logic **C** of Heinrich Wansing, and some basic results including soundness and completeness results are observed.

1 Introduction

In a recent paper [3], Nissim Francez introduces a system in which an unusual (even from connexivists' perspective!) formula $\sim(A{\to}B){\leftrightarrow}(\sim A{\to}B)$ is derivable. Francez's idea behind this formula is to express disagreement about the necessity of the succedent for the antecedent instead of the negated antecedent of a conditional. Although I must confess that I'm not very convinced about the motivation of Francez, the suggested formula itself is of great interest from a purely technical perspective, especially in view of an understanding of connexive logics suggested by Heinrich Wansing (see [6] for an application of Wansing's idea to the basic relevant logic **BD**). Based on these, the aim of this note is to examine the concerned formula in the light of Nelson's logic **N4** (cf. [7, 4]) by introducing a variant of **N4**, and compare this system with the connexive logic **C** introduced by Wansing in [8].

---

*The author is a Postdoctoral Research Fellow of the Japan Society for the Promotion of Science (JSPS). I would like to thank Heinrich Wansing for his encouragement as well as his helpful suggestions and comments, and Thomas Ferguson for some helpful suggestions.

2 Semantics and proof theory

The language $\mathcal{L}$ consists of a finite set $\{\sim, \wedge, \vee, \rightarrow\}$ of propositional connectives and a countable set Prop of propositional variables which we denote by p, q, etc. Furthermore, we denote by Form the set of formulas defined as usual in $\mathcal{L}$. We denote a formula of $\mathcal{L}$ by A, B, C, etc. and a set of formulas of $\mathcal{L}$ by Γ, Δ, Σ, etc.

2.1 Semantics

The following semantics is obtained by making a simple change to the standard semantics for Nelson's logic **N4**.

Definition 1. A model for the language $\mathcal{L}$ is a triple $\langle W, \leq, V \rangle$, where W is a non-empty set (of states); $\leq$ is a partial order on W; and $V : W \times \mathsf{Prop} \longrightarrow \{\emptyset, \{0\}, \{1\}, \{0,1\}\}$ is an assignment of truth values to state-variable pairs with the condition that $i \in V(w_1, p)$ and $w_1 \leq w_2$ only if $i \in V(w_2, p)$ for all $p \in \mathsf{Prop}$, all $w_1, w_2 \in W$ and $i \in \{0, 1\}$. Valuations V are then extended to interpretations I to state-formula pairs by the following conditions:

- $I(w, p) = V(w, p)$,
- $1 \in I(w, \sim A)$ iff $0 \in I(w, A)$,
- $0 \in I(w, \sim A)$ iff $1 \in I(w, A)$,
- $1 \in I(w, A \wedge B)$ iff $1 \in I(w, A)$ and $1 \in I(w, B)$,
- $0 \in I(w, A \wedge B)$ iff $0 \in I(w, A)$ or $0 \in I(w, B)$,
- $1 \in I(w, A \vee B)$ iff $1 \in I(w, A)$ or $1 \in I(w, B)$,
- $0 \in I(w, A \vee B)$ iff $0 \in I(w, A)$ and $0 \in I(w, B)$,
- $1 \in I(w, A \rightarrow B)$ iff for all $x \in W$: if $w \leq x$ and $1 \in I(x, A)$ then $1 \in I(x, B)$,
- $0 \in I(w, A \rightarrow B)$ iff for all $x \in W$: if $w \leq x$ and $0 \in I(x, A)$ then $1 \in I(x, B)$.

Finally, the semantic consequence is now defined as follows: $\Sigma \models A$ iff for all models $\langle W, \leq, I \rangle$, and for all $w \in W$: $1 \in I(w, A)$ if $1 \in I(w, B)$ for all $B \in \Sigma$.

Remark 2. Note that Nelson's logic **N4** is obtained by replacing the falsity condition for implication by the following condition.

$$0 \in I(w, A \rightarrow B) \text{ iff } 1 \in I(w, A) \text{ and } 0 \in I(w, B).$$

Moreover, Wansing's connexive logic **C** is obtained by replacing the falsity condition for implication by the following condition.

$$0 \in I(w, A \rightarrow B) \text{ iff for all } x \in W: \text{ if } w \leq x \text{ and } 1 \in I(x, A) \text{ then } 0 \in I(x, B).$$

2.2   Proof Theory

We now turn to the proof theory. Since Wansing's connexive logic $\mathbf{C}$ is presented in terms of a Hilbert-style calculus, we follow the same strategy.

Definition 3. The system $\mathcal{N}$ consists of the following axiom schemata and a rule of inference where $A \leftrightarrow B$ abbreviates $(A \to B) \wedge (B \to A)$:

(Ax1) $\qquad A \to (B \to A)$

(Ax2) $(A \to (B \to C)) \to ((A \to B) \to (A \to C))$

(Ax3) $\qquad\qquad (A \wedge B) \to A$

(Ax4) $\qquad\qquad (A \wedge B) \to B$

(Ax5) $(C \to A) \to ((C \to B) \to (C \to (A \wedge B)))$

(Ax6) $\qquad\qquad A \to (A \vee B)$

(Ax7) $\qquad\qquad B \to (A \vee B)$

(Ax8) $(A \to C) \to ((B \to C) \to ((A \vee B) \to C))$

(Ax9) $\qquad \sim\sim A \leftrightarrow A$

(Ax10) $\quad \sim(A \wedge B) \leftrightarrow (\sim A \vee \sim B)$

(Ax11) $\quad \sim(A \vee B) \leftrightarrow (\sim A \wedge \sim B)$

(Ax12) $\quad \sim(A \to B) \leftrightarrow (\sim A \to B)$

(MP) $\qquad \dfrac{A \quad A \to B}{B}$

Finally, we write $\Gamma \vdash A$ if there is a sequence of formulas $B_1, \ldots, B_n, A$, $n \geq 0$, such that every formula in the sequence $B_1, \ldots, B_n, A$ either (i) belongs to Γ; (ii) is an axiom of $\mathcal{N}$; (iii) is obtained by (MP) from formulas preceding it in sequence.

Remark 4. Note that if we replace (Ax12) by '$\sim(A \to B) \leftrightarrow (A \wedge \sim B)$', then we obtain an axiomatization of Nelson's logic $\mathbf{N4}$. Moreover, compared to the system $\mathbf{C}$ of Wansing, the only difference is again the axiom (Ax12). More specifically, (Ax12) is replaced by '$\sim(A \to B) \leftrightarrow (A \to \sim B)$'.

Before turning to the soundness and completeness proofs, we note that the deduction theorem is provable.

Proposition 1. *For any $\Gamma \cup \{A, B\} \subseteq$ Form, $\Gamma, A \vdash B$ iff $\Gamma \vdash A \to B$.*

Proof. It can be proved in the usual manner in the presence of axioms (Ax1) and (Ax2), given that (MP) is the sole rule of inference. $\qquad \square$

3   Soundness and completeness

As usual, the soundness part is rather straightforward.

Theorem 1 (Soundness). *For $\Gamma \cup \{A\} \subseteq$ Form, if $\Gamma \vdash A$ then $\Gamma \models A$.*

Proof. By induction on the length of the proof. $\qquad \square$

For the completeness proof, we first introduce some standard notions.

Definition 5. A set of formulas, Σ, is *deductively closed* iff if $\Sigma \vdash A$ then $A \in \Sigma$. And Σ is *prime* iff $A \vee B \in \Sigma$ implies $A \in \Sigma$ or $B \in \Sigma$. Moreover, Σ is *prime deductively closed* (pdc) if it is both. Finally, Σ is *non-trivial* if $A \notin \Sigma$ for some A.

The following two lemmas are well-known, and thus the proofs are omitted. We only note in passing that the deduction theorem is the key for the second lemma.

Lemma 1. *If $\Sigma \nvdash A$ then there is a non-trivial pdc, Δ, such that $\Sigma \subseteq \Delta$ and $\Delta \nvdash A$.*

Lemma 2. *If Σ is pdc and $A{\to}B \notin \Sigma$, there is a non-trivial pdc Θ such that $\Sigma \subseteq \Theta$, $A \in \Theta$ and $B \notin \Theta$.*

Now, we are ready to prove the completeness.

Theorem 2 (Completeness). *For $\Gamma \cup \{A\} \subseteq \mathsf{Form}$, if $\Gamma \models A$ then $\Gamma \vdash A$.*

Proof. We prove the contrapositive. Suppose that $\Gamma \nvdash A$. Then by Lemma 1, there is a $\Pi \supseteq \Gamma$ such that Π is a pdc and $A \notin \Pi$. Define the model $\mathfrak{A} = \langle X, \leq, I \rangle$, where $X = \{\Delta : \Delta$ is a non-trivial pdc$\}$, $\Delta \leq \Sigma$ iff $\Delta \subseteq \Sigma$ and I is defined thus. For every state, Σ and propositional parameter, p:

$$1 \in I(\Sigma, p) \text{ iff } p \in \Sigma \text{ and } 0 \in I(\Sigma, p) \text{ iff } \sim p \in \Sigma$$

We show that this condition holds for any arbitrary formula, B:

$$(*) \qquad 1 \in I(\Sigma, B) \text{ iff } B \in \Sigma \text{ and } 0 \in I(\Sigma, B) \text{ iff } \sim B \in \Sigma$$

It then follows that $\mathfrak{A}$ is a counter-model for the inference, and hence that $\Gamma \nvDash A$. The proof of $(*)$ is by a simultaneous induction on the complexity of B with respect to the positive and the negative clause.
For negation: We begin with the positive clause.

$$1 \in I(\Sigma, \sim C) \text{ iff } 0 \in I(\Sigma, C)$$
$$\text{iff } \sim C \in \Sigma \qquad\qquad \text{IH}$$

The negative clause is also straightforward.

$$0 \in I(\Sigma, \sim C) \text{ iff } 1 \in I(\Sigma, C)$$
$$\text{iff } C \in \Sigma \qquad\qquad \text{IH}$$
$$\text{iff } \sim\sim C \in \Sigma \qquad\qquad \text{(Ax9)}$$

For disjunction: We begin with the positive clause.

$1 \in I(\Sigma, C \vee D)$ iff $1 \in I(\Sigma, C)$ or $1 \in I(\Sigma, D)$

iff $C \in \Sigma$ or $D \in \Sigma$ IH

iff $C \vee D \in \Sigma$ Σ is a prime theory

The negative clause is also straightforward.

$0 \in I(\Sigma, C \vee D)$ iff $0 \in I(\Sigma, C)$ and $0 \in I(\Sigma, D)$

iff $\sim C \in \Sigma$ and $\sim D \in \Sigma$ IH

iff $\sim C \wedge \sim D \in \Sigma$ Σ is a theory

iff $\sim(C \vee D) \in \Sigma$ (Ax11)

For conjunction: Similar to the case for disjunction.

For implication: We begin with the positive clause.

$1 \in I(\Sigma, C \rightarrow D)$ iff for all Δ s.t. $\Sigma \subseteq \Delta$, if $1 \in I(\Delta, C)$ then $1 \in I(\Delta, D)$

iff for all Δ s.t. $\Sigma \subseteq \Delta$, if $C \in \Delta$ then $D \in \Delta$ IH

iff $C \rightarrow D \in \Sigma$ $(\star)$

For the last equivalence $(\star)$, assume $C \rightarrow D \in \Sigma$ and $C \in \Delta$ for any Δ such that $\Sigma \subseteq \Delta$. Then by $\Sigma \subseteq \Delta$ and $C \rightarrow D \in \Sigma$, we obtain $C \rightarrow D \in \Delta$. Therefore, we have $\Delta \vdash C \rightarrow D$, so by (MP), we obtain $\Delta \vdash D$, i.e. $D \in \Delta$, as desired. On the other hand, suppose $C \rightarrow D \notin \Sigma$. Then by Lemma 2, there is a $\Sigma' \supseteq \Sigma$ such that $C \in \Sigma'$, $D \notin \Sigma'$ and Σ' is a pdc. Furthermore, non-triviality of Σ' is obvious by $D \notin \Sigma'$.

As for the negative clause, it is similar to the positive case.

$0 \in I(\Sigma, C \rightarrow D)$ iff for all Δ s.t. $\Sigma \subseteq \Delta$, if $0 \in I(\Delta, C)$ then $1 \in I(\Delta, D)$

iff for all Δ s.t. $\Sigma \subseteq \Delta$, if $\sim C \in \Delta$ then $D \in \Delta$ IH

iff $\sim C \rightarrow D \in \Sigma$ (†)

iff $\sim(C \rightarrow D) \in \Sigma$ (Ax12)

For the equivalence (†), the proof runs exactly the same with the equivalence $(\star)$ in the positive case. Thus, we obtain the desired result. $\square$

4 Basic observations

Let us now briefly examine the system $\mathcal{N}$.

Proposition 2. $\mathcal{N}$ *is inconsistent. That is, both $A{\to}(A{\to}A)$ and $\sim(A{\to}(A{\to}A))$ are provable in $\mathcal{N}$.*

Proof. Just note that the second formula is equivalent to $\sim A{\to}(A{\to}A)$. □

Remark 6. As noted by Wansing, **C** is also inconsistent at the level of propositional logic. For example, both $(A\wedge\sim A){\to}(A\vee\sim A)$ and its negation are provable in **C**.

Proposition 3. *The extension of $\mathcal{N}$ by $A\vee\sim A$ (LEM hereafter) is trivial.*

Proof. If we have LEM, then we have $((A{\to}(A{\to}A)){\to}B)\vee\sim((A{\to}(A{\to}A)){\to}B)$. In view of (Ax12), we obtain $((A{\to}(A{\to}A)){\to}B)\vee(\sim(A{\to}(A{\to}A)){\to}B)$. Since $A{\to}(A{\to}A)$ and $\sim(A{\to}(A{\to}A))$ are both provable in $\mathcal{N}$ by the previous proposition, we obtain $B\vee B$, and thus B. Therefore, the extension under concern is trivial. □

Remark 7. Compare this with **N4** and **C**. In the former case, the addition of LEM results in a three-valued logic known as **CLuNs** (without bottom element) in the literature (cf. [2] for a detailed study). In particular, the constructive implication collapses into the classical material implication. In the latter case, we obtain an intermediate logic with a connexive flavor, and this remains to be explored.

Proposition 4. *The extension of $\mathcal{N}$ by Peirce's law, i.e. $((A{\to}B){\to}A){\to}A$, is sound and complete with respect to the semantics induced by the following matrix with **t** and **b** as designated values.*

A	$\sim A$	$A\wedge B$	t	b	n	f	$A\vee B$	t	b	n	f	$A{\to}B$	t	b	n	f
t	f	t	t	b	n	f	t	t	t	t	t	t	b	b	f	f
b	b	b	b	b	f	f	b	t	b	t	b	b	b	b	n	n
n	n	n	n	f	n	f	n	t	t	n	n	n	b	b	b	b
f	t	f	f	f	f	f	f	t	b	n	f	f	b	b	t	t

Proof. Just consider the model with only one state. □

Remark 8. Compare this again with **N4** and **C**. In the former case, the addition of Peirce's law results in a four-valued logic, called **HBe** in [1], induced by the matrix obtained by replacing the truth table for implication as follows:

$A{\to}B$	t	b	n	f
t	t	b	n	f
b	t	b	n	f
n	t	t	t	t
f	t	t	t	t

In the latter case, the addition of Peirce's law results in a four-valued logic, called material connexive logic in [9], induced by the matrix obtained by replacing the truth table for implication as follows:

$A{\to}B$	t	b	n	f
t	t	b	n	f
b	t	b	n	f
n	b	b	b	b
f	b	b	b	b

An expansion of the material connexive logic, obtained by adding the Boolean complement, is introduced and examined in [5].

Proposition 5. *Aristotle's theses are provable in $\mathcal{N}$, but Boethius' theses are not.*

Proof. For the Aristotle's theses, Just note that $\sim(A{\to}\sim A)$ and $\sim(\sim A{\to}A)$ are equivalent to $\sim A{\to}\sim A$ and $\sim\sim A{\to}A$ respectively. For the Boethius' theses, for the non-derivability of $(A{\to}B){\to}\sim(A{\to}\sim B)$, just assign **b** and **t** to A and B respectively in the above truth table, and for the non-derivability of $(A{\to}\sim B){\to}\sim(A{\to}B)$, assign **b** and **f** to A and B respectively in the above truth table. □

Remark 9. If one takes Boethius' theses to be indispensable for connexive logics, then $\mathcal{N}$ is disqualified to be a connexive logic in view of the above proposition. In other words, Boethius' theses are independent of Aristotle's theses in general. This is the reason why I referred to the axiom (Ax12), due to Francez, as half-connexive in the title of this note.

References

[1] A. Avron. Natural 3-valued logics–characterization and proof theory. *Journal of Symbolic Logic*, 56:276–294, 1991.

[2] D. Batens and K. De Clercq. A rich paraconsistent extension of full positive logic. *Logique et Analyse*, 185-188:227–257, 2004.

[3] N. Francez. Natural deduction for two connexive logics. *IFCoLog Journal of Logics and their Applications*, this issue.

[4] N. Kamide and H. Wansing. *Proof Theory of N4-related Paraconsistent Logics*. Studies in Logic, Vol. 54. College Publications, London, 2015.

[5] H. Omori. From paraconsistent logic to dialetheic logic. In H. Andreas and P. Verdée, editors, *Logical Studies of Paraconsistent Reasoning in Science and Mathematics*. Springer, forthcoming.

[6] H. Omori. A simple connexive extension of the basic relevant logic **BD**. *IFCoLog Journal of Logics and their Applications*, this issue.

[7] H. Wansing. Negation. In L. Goble, editor, *The Blackwell Guide to Philosophical Logic*, pages 415–436. Basil Blackwell Publishers, Cambridge, MA, 2001.

[8] H. Wansing. Connexive modal logic. In R. Schmidt, I. Pratt-Hartmann, M. Reynolds, and H. Wansing, editors, *Advances in Modal Logic. Volume 5*, pages 367–383. King's College Publications, 2005.

[9] H. Wansing. Connexive logic. In E. N. Zalta, editor, *The Stanford Encyclopedia of Philosophy*. http://plato.stanford.edu/archives/fall2014/entries/logic-connexive/, Fall 2014 edition, 2014.

Received May 2016

www.ingramcontent.com/pod-product-compliance
Lightning Source LLC
Chambersburg PA
CBHW080659110426
42739CB00034B/3339